Risk Analysis, Hazard Mitigation and Safety and Security Engineering XIII

WIT*PRESS*

WIT Press publishes leading books in Science and Technology.
Visit our website for new and current list of titles.
www.witpress.com

WITeLibrary

Home of the Transactions of the Wessex Institute.
Papers published in this volume are archived in the WIT eLibrary in volume 214 of
WIT Transactions on The Built Environment (ISSN 1743-3509).
The WIT eLibrary provides the international scientific community with immediate and
permanent access to individual papers presented at WIT conferences.
Visit the WIT eLibrary at www.witpress.com/elibrary.

Thirteenth International Conference on Risk Analysis, Hazard Mitigation and Safety and Security Engineering

RISK/SAFE 2022

Honorary Chairman

Massimo Guarascio
University of Rome, "La Sapienza", Italy

Conference Chairmen

Santiago Hernández
University of A Coruña, Spain
Member of WIT Board of Directors

Fabio Garzia
University of Rome "La Sapienza", Italy

Mara Lombardi
University of Rome "La Sapienza", Italy

Andrea Fabbri
University of Milano-Bicocca, Italy

International Scientific Advisory Committee

Andrea Antonucci
Warren Axelrod
Marco Baldi
Michael Beer
Marco Bietresato
Richard Carranza
Galina Chebotareva
Carlos Cuadra
Gianluca Dini
Gerardus Janszen
Zhen-Gang Ji
Elena Magaril
David Novelo-Casanova
Roberto Perruzza
Antonella Pontrandolfi
Elena Rada
Genserik Reniers
Francesco Russo
Seddik Sakji
Daniel Santos-Reyes
Amaranath Sena Kumara
Michel Olivier Sturtzer
Sophie Trelat

Sponsored by

WIT Transactions on the Built Environment

International Journal of Computational Methods and Experimental Measurements

Organised by

Wessex Institute, UK
University of Rome "La Sapienza", Italy
University of Milano-Bicocca, Italy

WIT Transactions

Wessex Institute
Ashurst Lodge, Ashurst
Southampton SO40 7AA, UK

We would like to express thanks to all the conference Chairs and members of the International Scientific Advisory Committees for their efforts during the 2022 conference season.

Conference Chairs

International Scientific Advisory Committee Members 2022

Risk Analysis, Hazard Mitigation and Safety and Security Engineering XIII

Editors

Santiago Hernández
University of A Coruña, Spain
Member of WIT Board of Directors

Fabio Garzia
University of Rome "La Sapienza", Italy

Mara Lombardi
University of Rome "La Sapienza", Italy

Andrea Fabbri
University of Milano-Bicocca, Italy

WIT*PRESS* Southampton, Boston

Editors:

Santiago Hernández
University of A Coruña, Spain
Member of WIT Board of Directors

Fabio Garzia
University of Rome "La Sapienza", Italy

Mara Lombardi
University of Rome "La Sapienza", Italy

Andrea Fabbri
University of Milano-Bicocca, Italy

Published by

WIT Press
Ashurst Lodge, Ashurst, Southampton, SO40 7AA, UK
Tel: 44 (0) 238 029 3223; Fax: 44 (0) 238 029 2853
E-Mail: witpress@witpress.com
http://www.witpress.com

For USA, Canada and Mexico

Computational Mechanics International Inc
25 Bridge Street, Billerica, MA 01821, USA
Tel: 978 667 5841; Fax: 978 667 7582
E-Mail: infousa@witpress.com
http://www.witpress.com

British Library Cataloguing-in-Publication Data
A Catalogue record for this book is available
from the British Library

ISBN: 978-1-78466-483-1
eISBN: 978-1-78466-484-8
ISSN: 1746-4498 (print)
ISSN: 1743-3509 (on-line)

The texts of the papers in this volume were set individually by the authors or under their supervision. Only minor corrections to the text may have been carried out by the publisher.

Preface

This Volume contains the revised versions of selected papers presented at 13th Conference on Risk Analysis, Hazard Mitigation and Safety and Security Engineering, held online during 12–14 October 2022, organised by Wessex Institute in the UK, and University of Rome "La Sapienza" and University of Milano-Bicocca in Italy. The conference was sponsored by the WIT Transactions on the Built Environment. It was originally scheduled to take place in Rome, in Italy, but subsequently had to be held online due to political and military situation in Europe and some remaining steps of the COVID-19 pandemic.

This event is the merging of two previously independent conferences, namely, the International conference on Safety and Security Engineering that started in Rome in 2005 and the International conference on Risk Analysis and Hazard Mitigation that inaugurated in Valencia in 1998.

The purpose of the conference was to provide a forum for the presentation and discussion of the most recent research and industrial developments in the theoretical and practical aspects of safety and security engineering. This field, due to its special nature, is an interdisciplinary area of research and application that brings together, in a systematic way, many disciplines of engineering from the traditional to the most technologically advanced. The conference covered areas such as crisis management, security engineering, natural disasters and emergencies, terrorism, IT security, man-made hazards, pandemics, transportation security, protection and mitigation issues, among others. Other conference topics were also all aspects of risk management and hazard mitigation, associated with both natural and anthropogenic hazards. Current events help to emphasise the importance of the analysis and management of risk to planners and researchers around the world. Natural hazards such as floods, earthquakes, landslides, fires, epidemics, transportation, climate change, fake news and others have always affected human societies. The more recent emergence of the importance of man-made hazards is a consequence of the rapid technological advances made in the last few centuries. The interaction of natural and anthropogenic risks adds to the complexity of the problems.

The contributions contained in this book are distributed in several sections, entitled: Risk analysis, assessment and management, risk in construction and management, cybersecurity/e-security and safety and security engineering. The diversity of the fields and the quality of the contents makes this volume an important addition to the scientific literature in this field.

For this reason, the papers contained in this book, as well as those from previous conferences since 2000, have been archived in the eLibrary of the Wessex Institute (http://www.witpress.com/elibrary) where they are permanently accessible to the international scientific community.

The editors wishes to acknowledge the support of the authors, the members of the International Scientific Advisory Committee (ISAC), the referees, Marta Graczyk, the conference co-ordinator, as well as the WIT Press staff and Isabelle Rham, in particular.

Finally, the editors and ISAC members wish to honour the memory of the late Professor Carlos Brebbia, founder of Wessex Institute, who established this series of meetings having foreseeing its impact and appeal.

The Editors, 2022

Contents

Section 1: Risk analysis, assessment and management

Section 2: Risk in construction and management

Section 3: Cybersecurity/e-security

Section 4: Safety and security engineering

SECTION 1
RISK ANALYSIS, ASSESSMENT AND MANAGEMENT

CAN WE ASSESS LANDSLIDE HAZARDS IN THE VOLCANIC CRATER OF LAKE ALBANO, ROME, ITALY?

ANTONIO PATERA[1*] & ANDREA FABBRI[2†]
[1]Istituto Nazionale di Geofisica e Vulcanologia, Italy
[2]Dipartimento di Scienze dell'Ambiente e della Terra, Università di Milano-Bicocca, Italy

ABSTRACT

This study applies mathematical models for assessing landslide susceptibility around Lake Albano, a volcanic crater and resort area near the city of Rome, Italy. The hazards are mass movements of many different types, recorded for more than 2,100 years that continue occurring to date encroaching with expanding urbanization and socioeconomic activities. The study area surrounding the lake occupies 30 km^2, in the form of a digital raster of 1002 pixels × 1202 lines at 5 m resolution: 975,093 above the water level and 229,311 below it. Of those, 8,867 pixels indicate the location of 150 sub-aerial landslides and 34,028 pixels that of 65 sub-aqueous landslides, respectively, that is, high densities of mass movements. A database collected the most available information on the landslides: distributions, types, linear and polygonal forms, and sub-aerial or sub-aqueous locations. Digitized categorical maps of land use classes and lithology units, in addition to a continuous field of high-resolution topographic elevation data, represented their physical settings. From a dense grid of elevation points, continuous-value maps at 5 m resolution were the following: aspect, digital elevation model, slope, curvature, planform, and profile. The results of prediction modelling by a fuzzy set membership function and a logistic discriminant function were digital images ranking the study area into relative levels of susceptibility. The spatial support of the settings varied with landslide types and physiographic conditions. The levels integrated empirical likelihood values representing the contrast in settings for all the pixels in the presence of the landslides with the pixel in their absence for each landslide type within the study area. Such ranks tend to overlap in predictions from the two models and for different types of landslides. Predicting landslide susceptibility for the area is feasible and with low uncertainty; however, the volcanic and socioeconomic context is a main challenge to measures of hazard and risk avoidance.
Keywords: landslide susceptibility, volcanic crater, fuzzy sets, logistic discriminant functions, spatial support, prediction modelling.

1 INTRODUCTION

Lake Albano has represented a focus of attention since prehistoric times. The lake, located approximately 20 km from the metropolitan centre of Rome, has a volcanic origin. Geologically it belongs to the Quaternary volcanic district of the Colli Albani (Albani Hills). The volcanic setting hosting the lake shows inward inclined flanks and steep slopes over the internal walls of the crater where a multitude of slope failures have been and are taking place.

Due to its natural beauty and the vicinity to the city a variety of recreational activities and urbanizations have developed along the internal slopes of the crater and the edge of the coastline. They represent elements exposed not only to the high risk of slope instability but also to tsunamis waves as secondary effects of sub-aqueous landslides. Patera and Fabbri [1], provided a synthesis of the geology and geomorphology settings of the Colli Albani and Lake Albano. This is summarised here.

Hydrothermal circulation, seismicity, and uplifting are signs of volcanic activity in the volcanic complex of the Colli Albani. What is left of three eruptive phases of 600,000–350,000 years ago, 350,000–270,000 years ago, and 200,000 years ago, respectively, is a

* *ORCID: https://orcid.org/0000-0001-7641-4689*
† *ORCID: https://orcid.org/0000-0002-9055-5960*

final hydro-magmatic phase to the present. The last eruptive centre appears to be Lake Albano that produced slag deposits of 36,000 years of age, recent deposits of mudslides of 5,800 years ago and the flood of 398 BC, possibly a consequence of the injection of hot fluids rich in CO_2 [2]. Furthermore, many emissions of gas, H_2S and CO_2, are a main cause of natural hazard associated to the volcanic activity [3], [4].

The edge of the crater reaches various elevations ranging from 585 m to a minimum of 390 m.a.s.l. while the coastline is at approximately 290 m.a.s.l. Depth of the lake is 167 m and its surface area is about 6 km^2. The lake sediments completely cover the lake bottom with thickness up to 14 m. Observed were hiatus in sediments, variation of $CaCO_3$ deposition rates, lake level fluctuations, uplift of the soil and paroxysmal degassing phenomena. Within the crater, three large platforms were identified as follows. The first forms the actual beach, and reaches 10 to 20 m below the water level interrupted by an 80 m escarpment. The second is ring shaped at the centre-south of the lake leading to a second escarpment of 35–50 m. The third is at the bottom of the lake at 126–133 m.a.s.l. [5].

The presence of a potentially active volcano in a densely populated area like the metropolitan Rome has led various researchers to evaluate volcanic risk in a multitude of studies on monitoring the gases dissolved in the lake, landslide risk in the internal slopes of Lake Albano including tsunamis risk of particular impact on an enclosed basin [5]–[7].

To account for the continued slope failure activity at Lake Albano, a simple Web query, such as "frane sul Lago Albano" (landslides on Lake Albano), returns pieces of news on local media on landslides just occurred on the following dates: 21 February 2011, 22 March 2011, 30 November 2014, 23 February 2015, 22 December 2015, 28 January 2016, 14 September 2017, 14 January 2018 and 28 September 2020. In addition, a query like "landslides on Lake Albano" returns a multitude of scientific papers about landslide studies concerning the lake. Furthermore, various social and cultural activities are taking place concerning Lake Albano and its protection. A list and description of their websites is in the Appendix. It testifies to the high degree of socioeconomic and environmental concern of the local communities.

Encouraged by the visible scientific and social interest in studying and protecting the lake area and the communities living around the lake, or touring it to enjoy its amenities, we have collected available data constructing a digital database. Its purpose is spatial prediction modelling of landslide susceptibility in the Lake Albano surroundings. Following a pilot study to document the database, its quality and suitability for predictive modelling [1], our analysis attempts to provide interpretable and robust assessments of a high density area of landslides occurrences in both the sub-aerial and the sub-aqueous settings. With a high density of events and a variety of dynamic types of landslides, can we provide acceptable and useable assessments?

To provide answers to this question, the following sections describe the database, introduce concepts and models, and discuss experiments leading to *prediction patterns* expressing the likelihood or possibility of future landslide occurrences estimated from the available data. We finally point out the limitations and complexities of the task.

2 THE DATABASE

The Lake Albano study area covers 30 km^2. Fig. 1 shows its location in the Lazio Region. The distribution of 149 sub-aerial landslide forms and 65 sub-aqueous forms is in Table 1(a) and Fig. 2, where linear and polygonal forms are grouped into dynamic types in the two physical settings, above and below the water level of the lake shown as a blue line. In the illustration, they overlay an enhanced digital elevation model. The three platforms and the two escarpments are well visible in the model providing the physical context of the sub-

Figure 1: Location of the study area around Lake Albano near Rome, Lazio Region, Italy.

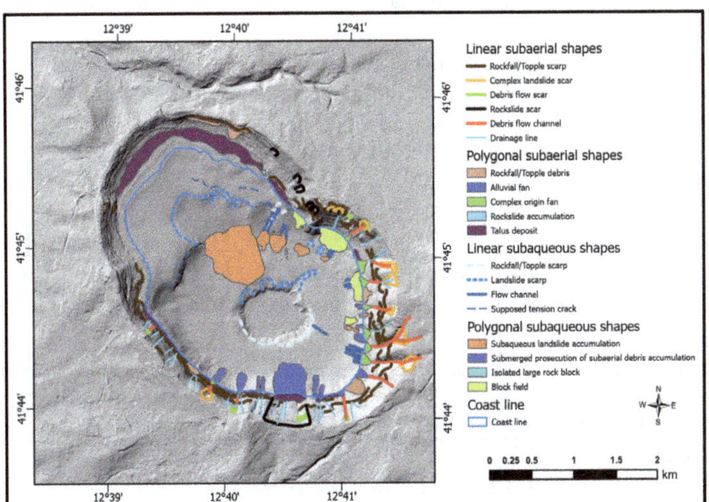

Figure 2: Map distribution of landslide forms of different dynamic types over the internal slopes of Lake Albano in (a), modified after [1].

aqueous forms. A southeast looking 3D view of the lake and its surrounding can be seen in Fig. 3.

The digital images of the individual groups of landslide forms, shown in Fig. 2, were instrumental in representing their distribution within the physical context of other digitized

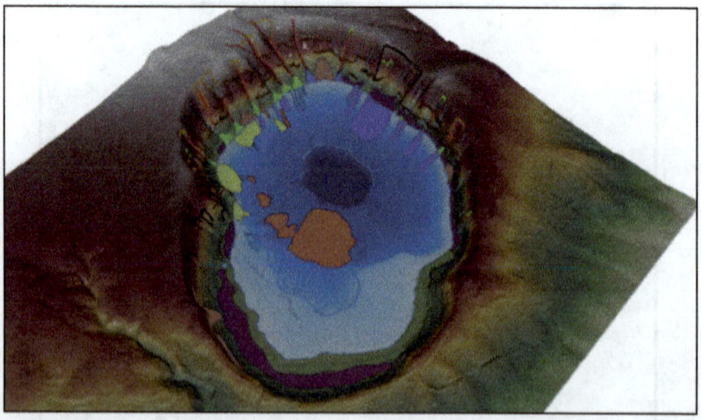

Figure 3: Southeast-looking 3D view of Lake Albano and the landslide forms.

maps: land use, lithology, digital elevation model, and its derivative continuous fields of aspect, slope, curvature, planform, and profile, listed in Table 1(b). Eleven classes of land uses and six lithological units were available from public sources. However, these were heterogeneous, of diverse periods, of different resolutions and poorly compatible with maps of slope failure processes. For those reasons, land use and lithology may not have a sufficient or appropriate resolution not only in spatial terms but also in the type and number of classes or map units. Finer and more focused mappings, although desirable, were not available to date [1].

The abbreviations in column 2 of Table 1(b) were used in various experiments to indicate the inputs in each experiment. Minimum and maximum values, in columns 3 and 4, indicate the numerical ranges of each continuous field. The description of each ISP, in column 5, indicates the implied relationships with the landslide processes.

Table 1: (a) Landslide forms and short names of the Lake Albano study area classified according to Cruden and Varnes [8] and Mulder and Cochonat [9]. They were termed direct supporting patterns (DSP). (b) Digitized categorical land use and lithology maps and DTM-derived continuous field maps with value ranges. They were termed indirect supporting patterns (ISPs).

(a)			
Sub-aerial linear forms		**Sub-aquatic linear forms**	
Debris flow channel	**LT1**	Flow channel	**LA1**
Debris flow scar	**LT2**	Landslide scarp	**LA2**
Drainage line	**LT3**	Rockfall/topple scarp	**LA3**
Rockfall/topple scarp	**LT4**	Supposed tension crack	**LA4**
Complex landslide scar	**LT5**		
Rockslide scar	**LT6**		
Sub-aerial polygonal forms		**Sub-aquatic polygonal forms**	
Alluvial fan	**PT1**	Block field	**PA1**
Complex origin fan	**PT2**	Isolated large rock block	**PA2**
Rockfall/Topple debris	**PT3**	Sub-aqueous landslide accumulation	**PA3**
Rockslide accumulation	**PT4**	Submerged prosecution of sub-aerial accumulation	**PA4**
Talus deposit	**PT5**		

Table 1: Continued.

(b)				
Pattern	**Short name**	**Minimum value**	**Maximum value**	**Description**
Land use	**U**	1	11	**U_1**, Wooded areas; **U_2**, Mining areas, construction sites, landfills; **U_3**, Urbanized green areas; **U_4**, Permanent crops; **U_5**, Shrubs or herbaceous vegetation cover; **U_6**, Productive settlement; **U_7**, Residential settlement; **U_8**, Stable lawns (permanent forages); **U_9**, Arable land; **U_{10}**, Heterogeneous agricultural areas; **U_{11}**, Open areas with sparse or absent vegetation
Lithology	**L**	1	6	**L_1**, Breccia; **L_2**, Gravel/sand/clay; **L_3**, Silt/clay; **L_4**, Lava (leucititis/trachyte); **L_5**, Slag/lapilli; **L_6**, Tuff
Aspect	**a**	0	359 d	In degrees from the north direction
DTM (elevation)	**d**	121.95 m	636.15 m	In m.a.s.l.
Slope	**s**	0 d	76.04 d	In degrees
Curvature	**c**	−236.58	241.66	Planform + Profile
Planform	**f**	−151.60	149.05	Divergence and convergence of the water flow (+ convex surface; − concave surface in direction of maximum slope)
Profile	**p**	−135.87	117.87	Acceleration and deceleration of the water flow (+ concave surface; − convex surface perpendicularly to the direction of maximum slope)

Fig. 4 shows the database of raster images. They were digitized at a resolution of 5 m × 5 m within a raster of 1002 pixels × 1202 lines (= 1,204,404 pixels), all in pixel-to-pixel correspondence. Within such raster, 975,093 pixels occupy the sub-aerial (terrestrial) part of the study area, while 229,311 pixels correspond to the sub-aqueous part.

In our analysis, the images of the presence of landslide forms were considered as direct supporting pattern, DSP, for the modelling. The remaining images were considered as indirect supporting patterns, ISPs, hopefully representing the typical setting of the landslides. Patera and Fabbri [1] constructed and discussed the database and its properties.

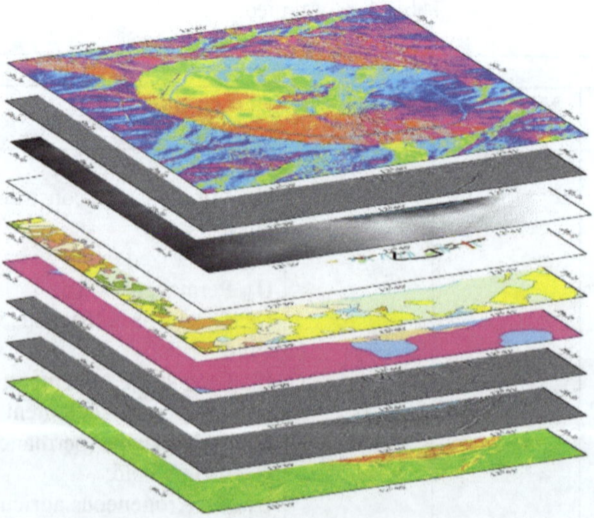

Figure 4: The database of digital images used for landslide susceptibility modelling in the Lake Albano study area, modified after [1].

Mathematical models of spatial prediction are to establish spatial relationships between DSP and ISPs and integrate them into *prediction patterns* representing the likelihood of future occurrences. For this, a unified mathematical framework and appropriate strategies are required. Chung and Fabbri [10] termed it "favourability function" modelling. Favourability modelling was applied to the database of Lake Albano. Following are terminology and strategies for the corresponding approach.

3 FAVOURABILITY FUNCTION MODELLING

The term "favourability" was proposed [10] to allow constructing functions for modelling the prediction of spatial patterns of future events given information on past events and their physical and possibly temporal settings. The mathematical framework of favourability function modelling is intended for applications in the fields of resource exploration and of natural hazard/risk assessment among others.

The applications could be made using different interpretations of the spatial relationships observed in areas where the hazardous events occurred (or resources discoveries were made). In essence, a favourability function is based on a proposition, i.e. a mathematical statement, to be proven true given the evidence available in the study area.

Because in our case we are dealing with landslides occurrences, representing the hazardous events, a modelling proposition is the following:

$$\mathbf{P}_i\text{: that a point } \boldsymbol{i} \text{ in the study area is affected by a specific type of occurrence} \mid \text{the presence of spatial evidence,} \tag{1}$$

where the symbol | indicates "given", the point ***i*** is a pixel of 5 m resolution, the specific type of occurrence is the presence of one of the landslide forms in Fig. 2 (the DSP), and the spatial evidence is the presence of one or more of the classes, units or values in the images in Fig. 4 (the ISPs).

To support the proposition in (1) we can explore the database, constructed by experts for the study area, and establish spatial relationships between the spatial evidence at the locations of the occurrences, DSP, and that at the locations of their absence. Obviously, we expect the study area to be suitable to provide the support sought, and the setting in the presence of the landslides to differ from that in their absence.

A common way to establish the spatial relationships between the presence and the absence in a study area is to calculate the empirical likelihood ratios or ELR. It consists of obtaining first the normalized frequencies of the DSP, the area proportion occupied by a given type of landslides, within the study area for each ISP. Some ISPs are the categorical maps partially representing their setting, e.g., land use and lithology for the sub-aerial forms. In addition, we calculate the corresponding density functions for the presence of the ISPs of continuous fields, e.g. aspect, DTM, slope, etc. We do the same also for the areas in which we assume the absence of the forms in the remainder of the study area. The ratio of the two normalized frequencies or density functions is the ELR, whose values range from zero to infinity. For instance, if for a pixel ELR is 2, it means that the frequency in the presence of the landslides is twice that in their absence in the study area. Should the ELR be equal to 1 it means that the two frequencies are the same in the presence of the landslides or in their absence in the study area. We tentatively interpret that ELR values ≥ 2.00 are supporting the modelling while values $\cong 1$ are not in support.

Establishing such spatial relationships is a first step in spatial modelling. There are many mathematical models for the integration of spatial relationships in favourability modelling, e.g. to measure the sureness or probability or certainty or belief, plausibility or possibility that the proposition in (1) is true. Chung and Fabbri [10] discussed Probabilistic, Certainty Factors, Dempster–Shafer Belief Function, and Fuzzy Logic Membership functions. Each interpretation and mathematical model requires a number of different assumptions.

Other assumptions are necessary concerning the database, the spatial relationships of DSP and the ISPs, and the strategy of processing. Fabbri and Chung [11] and Fabbri and Patera [12] discuss some of the assumptions in modelling landslide hazards. These are not going to be introduced here but mention will be made in the application examples. As a general procedure of predictive modelling we have selected the following steps:

1. Compute the spatial relationships and integrate them by a mathematical model;
2. Generate a prediction image from the integration and obtain from it a *prediction pattern* by sorting the values in descending order and equal-area ranking;
3. Cross-validate the *prediction pattern* to evaluate the quality of the prediction and generate prediction-rate curves for interpretation;
4. Estimate the uncertainty associated with the *prediction pattern* using a model and compare it to the uncertainty associated with patterns from different mathematical models.

In the applications that follow, we present the results of steps 1 to 3 and provide some anticipation of what to expect from a step 4. The two mathematical models applied are the Fuzzy Set function [10], [13], short named **FZ**, and the Logistic Regression Function [14], [15], short named **LO**.

4 EXAMPLE OF LANDSLIDE PREDICTION PATTERNS

The application example indicates how well we can answer the question in the title of this contribution. The procedure consists of the four steps introduced above. Spatial-relationship tables, *prediction patterns*, prediction-rate curves and overlays of top ranks are the modelling results for interpretation. To guide through steps and experiments, various short names indicate in sequence: the mathematical models, the direct supporting pattern, number of individual landslide occurrences and the sequence of indirect supporting patterns used in the analysis. The results of predictive modelling are the *prediction patterns*, i.e. all the pixels whose values are the possibility of future occurrences: for instance, the result of a Fuzzy Set function modelling for debris flow channel susceptibility in Table 1 is **FZ_LT1_9_ULads**.

4.1 The ELR tables

ELR values provide a relative measure of support for the modelling, i.e. for the proposition in (1). The DSP is the set of pixels with the presence of the landslides in the study area and the ELR values are a measure of contrast between the setting in the presence of the landslides and that in their absence in the study area.

Table 2 shows the sub-aerial landslide form names and short names in column 1, the number of individual landslides in column 2 and the corresponding number of pixels affected in column 3. Column 4 lists the ELR values for categorical ISPs. Column 5 shows the ELR values for continuous field ISPs. For the latter, the range of values above a threshold of ≥ 2 (sometime ≥ 1) are in italics font and within brackets is the maximum ELR reached and its corresponding value. All ELR values ≥ 2.00 are in bold fonts in the table. This simplification synthesizes histograms for categorical ISPs and functions for continuous-field ISPs for each landslide form.

For the polygonal forms, the number of pixels refers to the higher parts of approximately 15% of the area that represent the trigger zones. The terrestrial or sub-aerial part of the study area, visible in Fig. 2, covers 975,093 pixels.

Table 3 shows the ELR values for the sub-aqueous forms. No land use or lithology is available in this case but only the continuous field ISPs. The sub-aqueous part of the study area occupies 229,311 pixels. Here too the ELR values ≥ 2.00 are in bold fonts.

For instance, in Table 2, the first form **LT1**, with 9 debris flow channels affecting 807 pixels, have ratios ≥ 2 for $\mathbf{U_9}$ and $\mathbf{L_1}$, **3.38** and **2.46**, respectively arable land and lava. Supporting ISPs are also **d**, DTM, with values between *453* and *575* m.a.s.l. with a maximum ratio of **6.79** at *541* m elevation; and **s**, slope, between *19* and *72* degrees with a maximum of **7.34** at *56* degrees.

For sub-aerial forms, 120 linear and 29 polygonal in Table 2, we can see that $\mathbf{U_9}$ and $\mathbf{L_4}$ have high ratios for all except **PT5**; $\mathbf{L_1}$ is supportive of **LT5, LT6, PT3, PT4** and **PT5**; $\mathbf{U_{10}}$ is supporting **LT3** and **LT6**. Such common high ratios indicate some similarity of setting of the different landslide forms.

The sub-aqueous forms in Table 3 show groups of landslides with high ratios for distinct elevations: *155* m (**LA2, LA3**), *178* m (**PA2**), *200* m (**LA1**), *217* m (**PA3**), *245–246* m (**LA4, PA4**) and *282* m (**PA1**); shallow slope angles *4–21* d and *8–9* d (**PA2, PA3**), *10* and *20* d (**LA2, LA1**). Except for **LA1** and **PA3**, all other forms have high ELR values for **c**, **f** and **p**. In particular, **PA4** shows wide ranges for **a** and **s**, aspect and slope, *99* and *42* degrees, respectively.

In essence, the tables providing ELR values for each landslide form describe the spatial signature of the landslides within the study area's sub-aerial and sub-aqueous parts. This is what the database can provide the mathematical models for prediction.

Necessary assumptions for the modelling are: that the database sufficiently represents the characteristics of the landslide process, that the ISPs provide factors related to it, that the distribution of the landslides is a sample of a larger population in space and in time, and that we can separate them into two groups, one for modelling and the other for verification being

Table 2: ELR values for linear, **LT**, and polygonal, **PT**, top 15% sub-aerial (terrestrial) forms, DSP, using as ISPs land use, **U**, lithology, **L**, and the continuous fields derived from the elevation, **a**, **d**, **s**, **c**, **f** and **p**. Abbreviations are as in Table 1. Values are bold if ELR ≥ 2.00. The corresponding ranges of classes are in italics with the maximum class and ratio in brackets.

Sub-aerial linear forms				
Forms	**No.**	**Pixels**	**Categorical ISP**	**Continuous ISP**
LT1 Debris flow channel	9	807	**U_9=3.38;** U_{10}=1.92; L_1=1.05; **L_4=2.46;** L_6=0.78	a ≥ 1 *145–211* (max 1.94 at *171*d), *245–330* (max 1.42 at *307*d); **d** ≥2 *453–575* (max **6.79** at *541*m); s≥2 *19–72* (max **7.34** at *56*d)
LT2 Debris flow scar	14	764	**U_9=3.76;** U_{10}=1.07; **L_1=2.26;** **L_4=2.44;** L_5=1.73; L_6=0.34	a ≥ 1 *177–244* (max 1.55 at *206*d), *285–337* (max 1.34 at *309*d); **d** ≥2 *479–496* (max **2.03** at *487*m); *528–569* (max **2.65** at *551*m); **s** ≥2 *22–70* (max **12.92** at *57*d)
LT3 Drainage line	7	177	**U_9=3.22;** **U_{10}=3.29;** L_1=1.41; L_4=1.64; L_6=0.91	**a** ≥2 *0–83* (max **3.76** at *14*d); *353–361* (max **2.30** at *361*d); **d** ≥2 *430–519* (max **4.54** at *469*m); **s** ≥2 *22–73* (max **82.39** at *73*d)
LT4 Rockfall topple scarp	33	1344	**U_9=4.01;** **L_1=10.72;** L_2= 0.18; **L_4=2.60;** **L_5=2.84;** L_6=0.23	**a** ≥2 *0–75* (max **3.89** at *16*d); **d** ≥1 *334–503* (max 1.95 at *449*m); **s** ≥2 *23–72* (max **11.50** at *46*d)
LT5 Complex landslide scar	51	2969	U_6=0.18; **U_9=3.44;** U_{10}=1.93; **L_1=9.15;** L_2=0.35; **L_4=3.29;** **L_5=6.40;** L_6=0.16	**a** ≥2 *25–84* (max **2.88** at *53*d); **d** ≥1 *305–414* (max 1.80 at *354*m); *426–484* (max 1.52 at *459*m); **s** ≥2 *25–75* (max **67.90** at *72*d)
LT6 Rockslide scar	6	500	U_1=0.19; **U_9=2.09;** **U_{10}=7.15;** **L_1=3.90;** **L_4=3.36;** L_6=0.46	**a** ≥2 *0–45* (max **2.70** at *29*d); **d** ≥2 *386–415* (max **2.21** at *400*m); *479–524* (max **3.86** at *502*m); **s** ≥2 *22–66* (max **9.79** at *38*d and max **4.69** at *60*d)

Table 2: Continued.

Sub-aerial polygonal forms				
Forms	**No.**	**Pixels**	**Categorical ISP**	**Continuous ISP**
PT1 Alluvial fan	2	95	**U_9=4.01; L_4=6.91**	**a** ≥2 *309–352* (max **3.09** at *351*d); **d** ≥2 *294–345* (max **4.90** at *320*m); **s** ≥2 *18–37* (max **8.20** at *24*d)
PT2 Complex origin fan	11	416	**U_9=3.81;** U_{10}=0.84; **L_1=2.04; L_2=16.83; L_4=3.72; L_5=6.38**	**a** ≥2 *262–298* (max **2.54** at *280*d); **d** ≥2 *292–345* (max **4.48** at *320*m); **s** ≥2 *20–68* (max **7.20** at *42*d)
PT3 Rockfall topple debris	6	543	**U_9=3.94;** U_{10}=0.31; **L_1=3.13; L_2=2.44; L_4=3.32;** L_6=0.44	**a** ≥2 *49–65* (max **2.30** at*57*d); *184–210* (max **2.60** at *196*d); **d** ≥2 *324–378* (max **2.72** at *350*m); **s** ≥2 *20–69* (max **7.85** at *34*d, max **8.25** at *62*d)
PT4 Rockslide accumulation	1	14	**U_6=2.42; U_9=2.29; L_1=24.99**	**a** ≥2 *200–249* (max **4.58** at *221*d); **d** ≥2 *274–328* (max **5.07** at *302*m); **s** ≥2 *14–27* (max **9.66** at *20*d)
PT5 Talus deposit	9	1883	U_1=0.48; U_6=1.13; **U_8=2.85;** U_9=1.46; U_{10}=1.06; **L_1=23.69;** L_2=0.70; L_4=0.34; **L_5=4.60**	**a** ≥2 *174–227* (max **3.71** at *198*d); **d** ≥2 *284–338* (max **4.75** at *313*m); **s** ≥1 *35–70* (max1 **4.56** at *51*d and max2 **7.98** at *69*d)

the two groups of comparable nature. In addition, the assumption made in generating the ELR values is that the part of the study area without landslide occurrences (or in which occurrences are unknown) shows sufficient contrast of signature for the modelling.

4.2 **FZ** and **LO** prediction patterns

The ELR values are the support that the database provides to the modelling. A *prediction pattern* is the result of applying a mathematical model for normalizing and integrating the ELR values into a relative favourability index to classify the study area, i.e. the prediction rates.

The *prediction patterns* of sub-aerial landslides in Fig. 5 were obtained applying the two models of Fuzzy Set function, **FZ** [13] (gamma function operator with $\gamma = 0.5$), and Logistic Discriminant function, **LO** [15]. The prediction rates of a model are the results of normalization and combination rules that differ from model to model. So that the resulting prediction image has values between a minimum and a maximum, not interpretable as such.

Table 3: ELR values for linear, **LA**, and polygonal, **PA**, higher 15% sub-aqueous forms, DSP, using as ISPs the continuous fields derived from the elevation, **a**, **d**, **s**, **c**, **f** and **p**. Abbreviations are as in Table 1. Values are bold if ELR ≥ 2.00. The corresponding ranges of classes are in italics with the maximum class and ratio in brackets.

Sub-aqueous linear forms			
Forms	**No.**	**Pixels**	**Continuous ISP**
LA1 Flow channel	21	578	**a** ≥ 2, *242–313* (max **5.03** at *285*d); c ≥ 1 *−43* to *−3* (max 1.14 at *−34*); **d** ≥ 2, *205–234* (max **2.39** at *217*m); f ≥ 1 *−26* to *−2* (max 1.15 at *−20*); p ≥ 1 *2–34* (max 1.48 at *28*); **s** ≥ 2, *0–76* (max **4.04** at *20*d)
LA2 Landslide scarp	10	1154	**a** ≥ 2, *154–183* (max **2.10** at *162*d); **c** ≥ 2, *215–222* (max **2.40** at 218); **d** ≥ 2, *146–166* (max **4.58** at *155*m); *238–261* (max **4.14** at *251*m); f ≥ 1, *139–150* (max 1.99 at *142*), *156–168* (max 1.50 at *162*); **p** ≥ 2, *148–153* (max **2.43** at *151*); s ≥ 1, *6–26* (max 1.92 at *10*d)
LA3 Rockfall/ topple scarp	5	444	**a** ≥ 2, *0–90* (max **2.23** at *0*d); **c** ≥ 2, *218–220* (max **2.03** at *219*); *254–274* (max **6.86** at *265*); **d** ≥ 2, *144–173* (max **10.08** at *155*); **f** ≥ 2, *135–144* (max **4.61** at *138*), *159–166* (max **3.46** at *163*); **p** ≥ 2, *113–128* (max **9.36** at *120*); **s** ≥ 2, *12–20* (max **2.95** at *16*d), *27–36* (max **3.04** at *32*d)
LA4 Supposed tension crack	6	456	**a** ≥ 2, *109–116* (max **2.03** at *112*d); *139–197* (max **3.51** at *166*d); **c** ≥ 2, *260–270* (max **2.16** at *266*); **d** ≥ 2, *230–268* (max **7.31** at *245*m); **f** ≥ 2, *164–165*; (max **2.03** at *165*); **p** ≥ 2, *118–125* (max **2.75** at *121*); **s** ≥ 2, *26–39* (max **6.68** at *33*d)
Sub-aqueous polygonal forms			
PA1 Block field	4	696	**a** ≥2 *194–285* (max **3.91** at *260*d); **c** ≥2 *−91* to *−16* (max **103.46** at *−47*); *+41* to *+80* (max **6.19** at *+66*); **d** ≥2 *267–290* (max **7.03** at *282*m); **f** ≥2 *−38* to *−8* (max **187.19** at *−38*); *+7* to *+20* (max **8.99** at *+13*); **p** ≥2 *−35* to *−19* (max **6.01** at *−29*); *+9* to *+50* (max **84.05** at *−26*); **s** ≥2 *26–63* (max **104.54** at *52*d)
PA2 Isolated rock block	3	21	**a** ≥2 *112–173* (max **4.23** at *144*d); c ≥1 *0* to *+18* (max 1.26 at *+13*); **d** ≥2 *168–188* (max **3.72** at *178*m); f ≥1 *0* to *+9* (max 1.35 at *+8*); p ≥ 1 *−11* to *+1* (max 1.08 at *−7*); **s** ≥2 *4–21* (max **2.10** at *4*d)
PA3 Sub-aqueous landslide accumulated	6	1231	**a** ≥2 *10–14* (max **2.65** at *12*d); c ≥1 *−1* to *+20* (max 1.12 at *+15*); **d** ≥2 *184–211* (max **5.34** at *200*m); f ≥1 *0* to *+10* (max 1.15 at *+7*); p ≥1 *−11* to *0* (max 1.07 at *−8*); **s** ≥2 *8–9* (max **2.05** at *9*d)
PA4 Submerged prosecution of sub-aerial accumulation	10	3461	**a** ≥2 *0–49* (max **4.85** at *0*d), *310–360* (max **7.16** at *360*d); **c** ≥2 *−75* to *−46* (max **2.57** at *−56*); **d** ≥2 *230–290* (max **2.84** at *246*m); **f** ≥2 *−45* to *−30* (max **2.86** at *−38*); **p** ≥2 *+25* to *+45* (max **4.47** at *+36*); **s** ≥2 *17–57* (max1 **4.60** at *25*d, max2 **6.15** at *42*d)

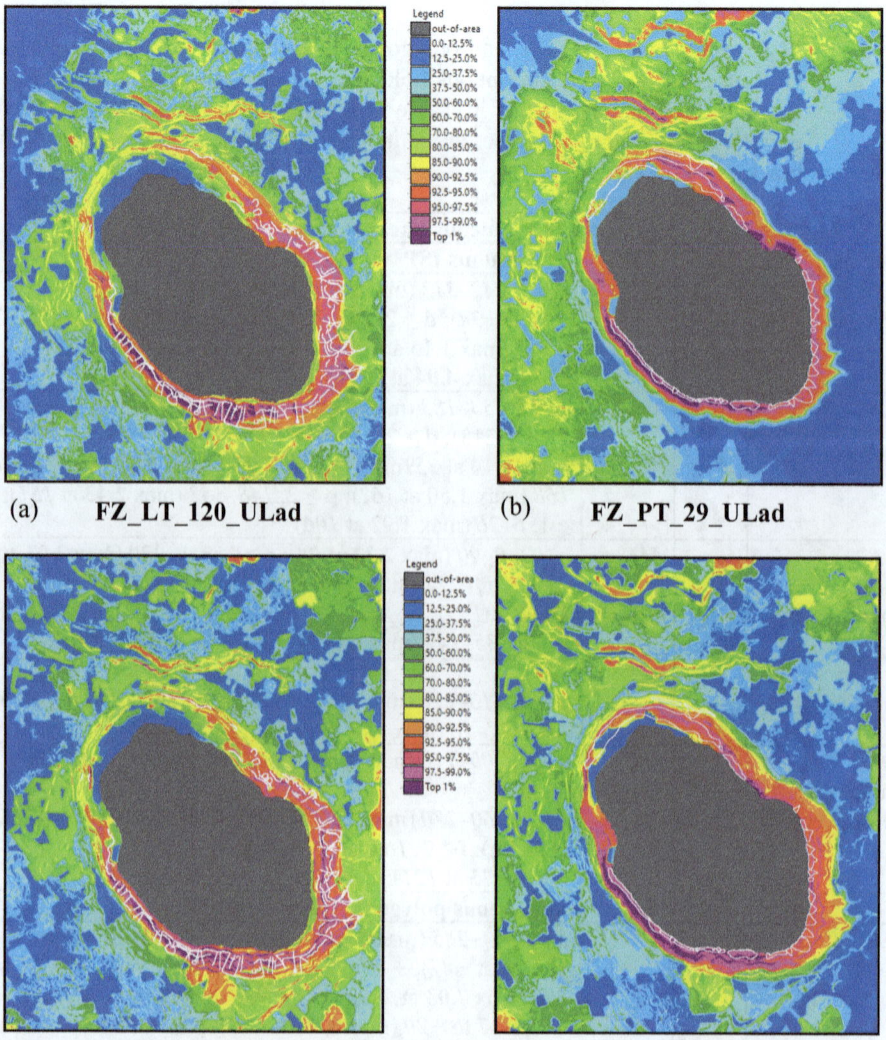

(a) **FZ_LT_120_ULad** (b) **FZ_PT_29_ULad**

(c) **LO_LT_120_ULad** (d) **LO_PT_29_ULad**

Figure 5: **FZ** (a) and (b), and **LO** (c) and (d) prediction patterns for linear and polygonal sub-aerial landslides, **LT_120** and **PT_29**, respectively.

By sorting these values in decreasing order and converting them into equal-area ranks we can visualize and interpret them as *prediction patterns*, i.e. spatial configurations of classes of recognizable shapes and locations.

In our case, we generated 200 equal-area ranks, each corresponding to 0.5% of the study area of 975,093 pixels, i.e. 4875 pixels. The ranks were then grouped into wider classes for the lower values and narrower classes for increasingly higher values of greater concern. Each class was then assigned a color from a pseudo-color look-up table according to the legend of the patterns. The legend was maintained the same for all patterns to allow recognition and comparison.

In the four illustrations of Fig. 5, the contours of the landslides used for modelling are superimposed as white lines. The classes from orange to purple, the top 10% of the study area, indicate the areas with the highest susceptibility to linear and polygonal sub-aerial landslide occurrences. In particular, the top 1% ranks, purple, indicate the critical areas and differences between patterns, useful for comparing **FZ** and **LO** predictions for either **LT** or **PT** landslides. There are similarities and differences to consider around the lake.

The density of the 120 **LT** landslides is high and the predicted areas of susceptibility concern are the ones far from the landslide occurrences in white in the patterns of Fig. 5. We can observe that the top 1% ranks are distributed along the southwestern coastline in Fig. 5(a) but in Fig. 5(c) they extend to the eastern part. Comparing the patterns in Fig. 5, we can see that the top 1% ranks extend to the northeastern parts in Fig. 5(b) while they are mostly on the southwestern coast in Fig. 5(d). The southwestern concentration of higher ranks indicates high susceptibility for both the linear and the polygonal sub-aerial landslides.

Assumptions combining the support of ISPs are of conditional independence of categorical, continuous fields and of their aggregation. Generally, in the earth science, maps are frequently conditionally dependent, and such dependence has to be verified to make sure it is not affecting the *prediction patterns*.

Furthermore, we need to know: how "good" the patterns are as predictors of "future" landslides of the same broad group of forms, linear or polygonal? This we attempt in the next section through cross-validation strategies.

4.3 Prediction-rate curves from cross-validations

Ideally, a prediction uses past and present event data to predict future events. However, we do not have sufficient information on the time of occurrence of the landslides in the Lake Albano area. What we can do is to partition a set of landslides into a modelling group and a cross-validation group, pretending that this second group represents the future landslide occurrences.

Cross-validation is the process of generating a *prediction pattern* using the modelling set and verifying the predicted values in the locations of the cross-validation set. The process can be iterative as sequential exclusion of a selected number of occurrences, as sequential selection, or as random selection repeated a convenient number of times [16].

Prediction-rate curves show on the horizontal axis the cumulative proportion of study area classified as susceptible, with ranks in decreasing order. On the vertical axis, they show the corresponding cumulative proportion of cross-validation occurrences.

In Fig. 6(a) and 6(b), the solid black curves are the results of iterative processes **FZ_LT_120m8_ULads** and **FZ_PT_29m2_Ulads**, where **m** indicates minus for exclusion of **8** and **2** occurrences. The curves are obtained by integrating the results of 15 and 14 iterations, respectively. The varicolored thinner curves are from the individual iterations where **8** or **2** occurrences were considered as the "next" ones.

The iterations in Fig. 6(a) show limited variability of curves that are all very steep and represent extremely good predictions. The top 5% ranks predict 60% of the occurrences and the top 10%, 87%, and top 15%, 99% on the thick black curve. In Fig. 6(b), the thick black curve shows a lower variability up to the top 5%. The top 5% predicts 45%, the top 10%, 53% and the top 15%, 58% of the occurrences. It is with iterations 9th and 13th that high deviations appear, i.e. lower prediction rates for the two corresponding pairs of landslides. Note the similarity of the solid red and blue curves in Fig. 6(c) with the solid black ones in Fig. 6(a) and 6(b). It means that the two models, **FZ** and **LO**, generate rather similar

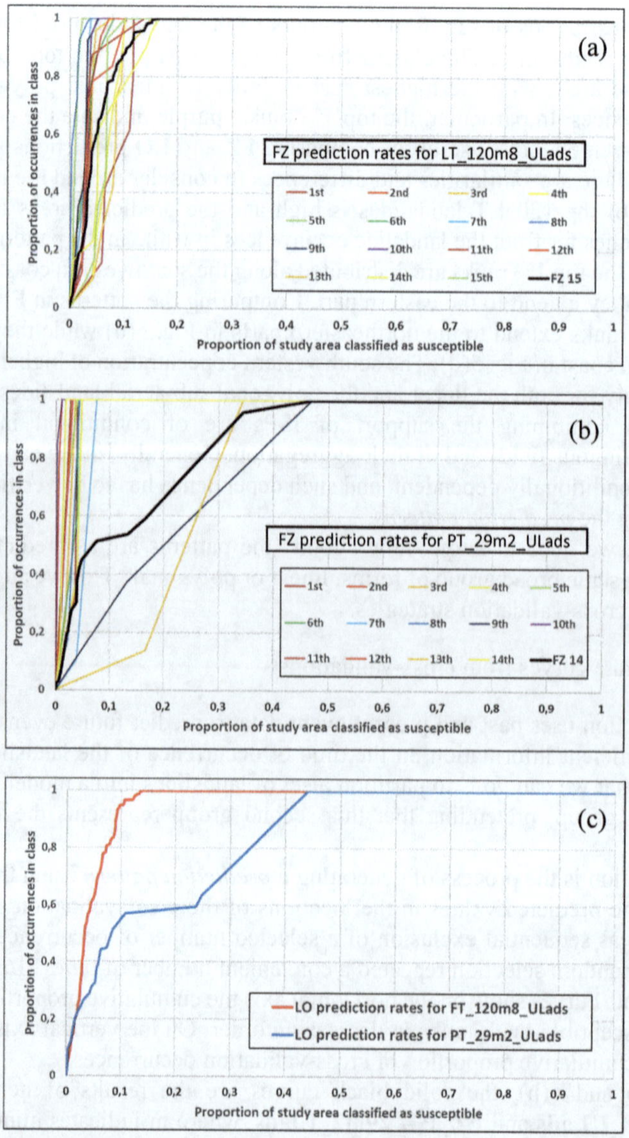

Figure 6: Prediction rate curves for the prediction patterns of linear and polygonal forms in Fig. 5. (a) The curves from iterative procedure **FZ_LT_120m8_ULads**; (b) The curves from **FZ_PT_29m2_ULads**; and (c) The curves from procedures **LO_LT_120m8_ULads** and **LO_PT_29m2_ULads**.

predictions. Altogether, having such limited variability in the prediction-rate curves means that the prediction patterns show low uncertainty of ranking.

Assumptions in the cross-validation process are that we can separate the landslides into two or more sets, younger and older to verify the consistency of *prediction patterns*, that we can assess the uncertainty associated with the patterns. To perform step 4, mentioned in Section 3, because the iterative cross-validation generated a prediction pattern per iteration, the set of patterns can be used for statistical analyses estimating the relative uncertainty associated with the initial pattern [12]. In our case the prediction-rate curves signal already a limited uncertainty

What are the significant parts of the prediction-rate curves and the corresponding top ranks in the *prediction patterns*? Can we compare the patterns for sub-aerial linear landslides with the polygonal ones? Can we compare the patterns with **FZ** models with those with **LO** models? This is done in the next section.

4.4 Comparison of top ranks of patterns

To visualize and interpret the degree of similarity of *prediction patterns*, a simple procedure was followed. The top 5% ranks from patterns **FZ_LT_120_ULads** and **FZ_PT_29_Ulads** were converted to numerical values of 1 while the remainder of the study area was converted to 0. Here we assume, tentatively, that the meaningful (or convenient) part of the prediction patterns is the top 5% ranks.

The corresponding binary, 0-1 images were overlaid with an operation of crossing to obtain a table of paired values, 0*0, 0*1, 1*0, and 1*1 with associated numbers of pixels, and a crossing image as shown in Fig. 7(a). The same procedure was applied to obtain Fig. 7(b) for the crossing of **FZ_LT_120_ULads** and **LO_LT_120_ULads**.

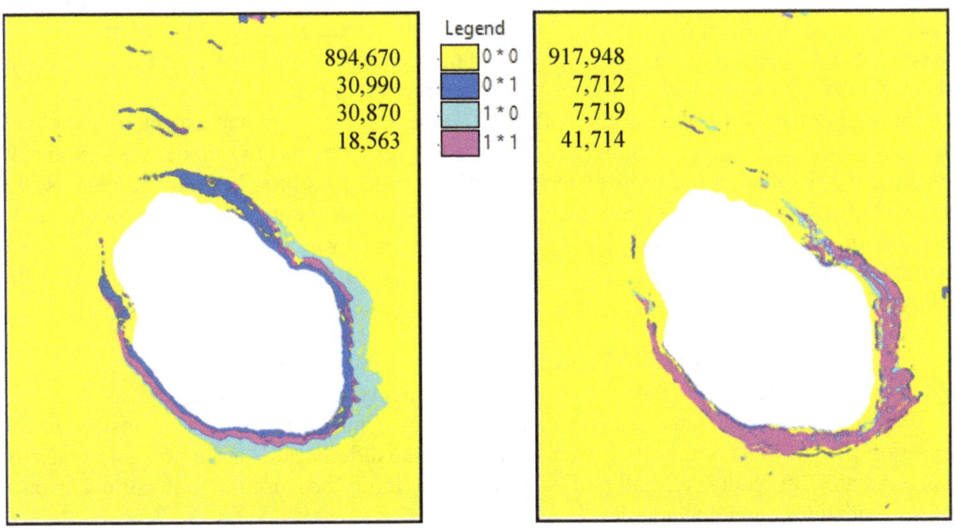

(a) **FZ_LT_120_ULads**
FZ_PT_29_ULads

(b) **FZ_LT_120_ULads**
LO_LT_120_ULads

Figure 7: Overlaps of the top 5% ranks of **FZ** and **LO** prediction patterns. (a) Overlaps of prediction patterns for **FZ** linear and polygonal forms; and (b) Overlaps of **FZ** and **LO** *prediction patterns* of linear forms. Legend indicates the number of overlapping top 5% ranking pixels.

The wide overlap, 1*1, of Fig. 7(b) shows that the corresponding *prediction patterns* are very similar and robust to changes in mathematical models. The limited overlap, 1*1, of Fig. 7(a), on the contrary, shows a limited but consistent overlap, 1*1, of higher susceptibility ranks for linear and/or polygonal sub-aerial landslides. They confirm the co-occurrence of susceptibilities to the different landslide forms, linear and polygonal.

Let us now consider a form of polygonal sub-aqueous landslides like Submerged prosecution of sub-aerial accumulation and generate *prediction patterns* in the sub-aqueous part of the study area, using the same two mathematical models.

4.5 Prediction of a polygonal sub-aqueous form

The sub-aerial forms **LT** and **PT** appear robust to changes in the mathematical model, from **FZ** to **LO**. In most cases in different studies the same robustness has been observed [11], [17]. However, for polygonal sub-aqueous form **PA4**, see the bottom of Table 3 for the ELR values, we observed a sensitivity to the models.

PA4 consists of 10 landslides, of which we later used their higher 15% pixels as DSP. We could only use **ads cfp** as ISPs and the sub-aqueous study area of 229,311 pixels. As can be observed in Table 3, all the 6 ISPs generate high ELR values, therefore all being supportive of the proposition in (1). The *prediction patterns* in Figs 8(a) and 7(b), **FZ_PA4_10_ads_cfp** and **LO_PA4_10_ads_cfp** show sharp differences. This in spite of the prediction rate curves from cross-validations processes (**FZ_PA4_10m1_ads_cfp** and **LO_PA4_10m1_ads_cfp**) generating very steep prediction-rate curves in Fig. 8(c). The top 5%, 10% and 15% ranks predict 50%, 95% and 100% of the occurrences for **LZ**, and 50%, 79% and 99% for **LO**. Fig. 8(d) shows the overlaps of their top 5% ranks (11,476 pixels) where 1*1 consists of 4,613 pixels while 0*1 and 1*0, consist of 7,000 and 6,976 pixels, respectively. In the illustration, the borders of the polygonal landslides have been plotted in black.

Observing the ELR values at the bottom of Table 3, we can say that aspect, **a**, and slope, **s**, have bimodal ELR functions that are given more weight by the **LO** model with extension of the top 5% ranks in the northeastern direction of the coastline. In both the models, the support of ISPs **c**, **f** and **p** is redundant. Prediction made without them appears identical to the ones using them being they highly correlated with elevation. While more can be done with the remaining sub-aqueous forms, analyzing the **PA4** landslides is a first step after the study of the sub-aerial ones.

5 CONSIDERATIONS AND CONCLUDING REMARKS

What are the limitations, complexities and improvements of the study and database?

Our analysis of the Lake Albano database have so far provided a clear answer to the question title of this contribution: Can we assess landslide hazard in the volcanic crater of Lake Albano? In reality, we had to deal with susceptibility because the times of occurrence of the landslides are not available.

The ELR tables provide clear support to favourability function modelling so that the *prediction patterns* of both the linear and the polygonal sub-aerial landslides are reliable representations of where future occurrences are to be expected. The sub-aerial patterns, given the similarities of settings and the high density of occurrences, had to be generated by grouping them into 120 linear and 29 polygonal landslides. The corresponding aggregated prediction-rate curves are very steep, "good predictions", and show very limited variability, which translates into low uncertainty of ranking.

For this reason, we have arbitrarily selected the top 5% ranks of the *prediction patterns* (48,750 pixels or ≅1.22 km^2 in the sub-aerial part of the study area) as the areas of relatively high susceptibility. Their overlaps from two different models, **FZ** and **LO**, and from the two groups of linear and polygonal landslides, show high degrees of similarities and robustness.

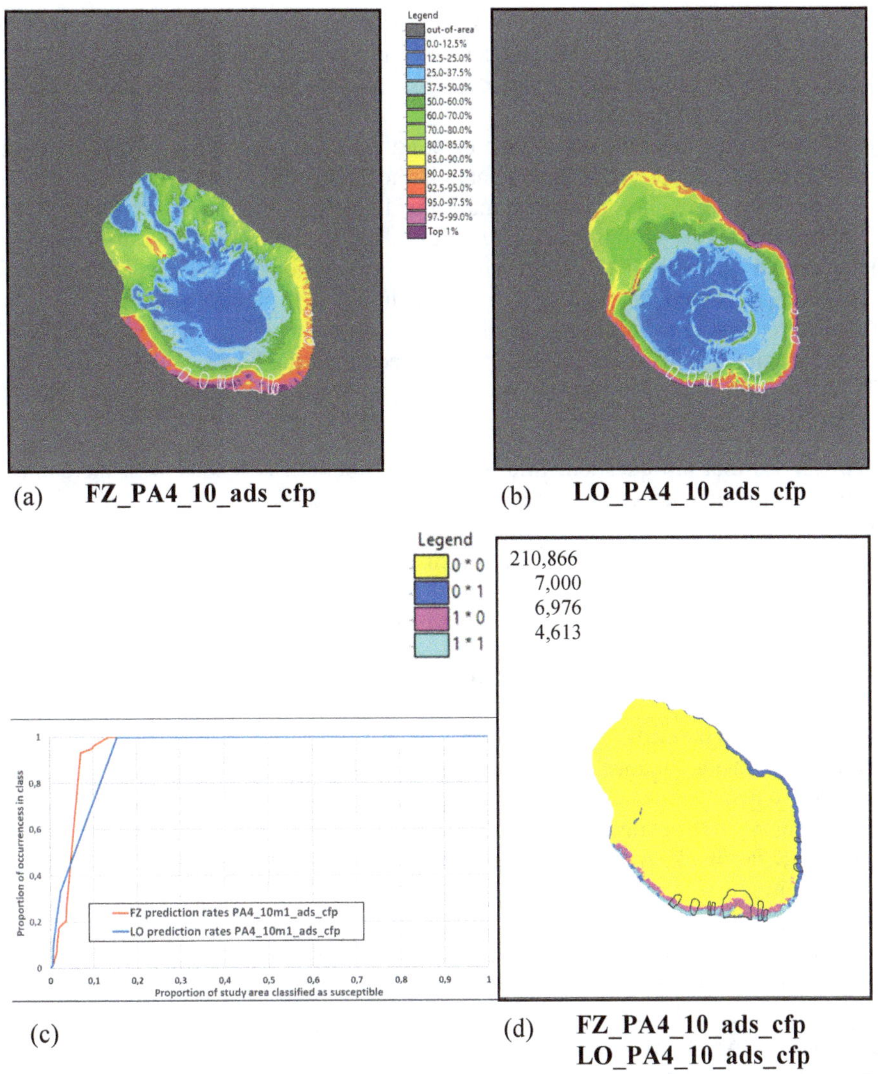

Figure 8: **FZ** and **LO** (a) and (b) prediction patterns for polygonal sub-aqueous landslides, FP4_10, contours in white. In (c) the corresponding 10m1 prediction-rate curves. In (d) the overlaps of the corresponding top 5% ranks of the two patterns, contours in black.

One example of prediction patterns from sub-aqueous polygonal landslides, while describing a particular setting of prosecution of sub-aerial accumulations, provides a particular instance of sensitivity to the two models due to wide spatial relationships of aspect and slope angles.

The following points are worth considering.

(i) Prediction of landslide susceptibility is very good: we know where the next sub-aerial occurrences will show up;

(ii) Anticipating the occurrences is good but the evidence is very wide: one should, in the future, use more appropriate and detailed land use and lithology maps;
(iii) Measures of mitigation of hazard or risk avoidance can be taken or applied with confidence over the patterns;
(iv) Much activity is taking place for protecting the Lake Albano's environment through a multitude of initiatives as documented in the Appendix;
(v) How could we interpret or estimate the likelihood that local administrations and communities are willing to invest in prevention, i.e. before the next disasters occur?
(vi) Could a more holistic approach consider additional multiple hazards such as volcanism, seismicity, and hydrologic conditions in the area?
(vii) Who could use the results of the assessments? Civil protection agencies, local authorities, public associations for the protection of the lake?
(viii) What is the societal convenience of acting in consequence of the assessment? Who benefits from the safeguard actions undertaken?

We can see that the question, title of this contribution, is loaded with incredulity and uncertainty, not on the technical feasibility of the task of obtaining reliable *prediction patterns* but on estimating the willingness to accept evaluations in order to manage such hazards and consequent risks of "future" events.

APPENDIX

Cultural activities

Compagnia dei viandanti https://compagniadeiviandanti.com/standby/anello-del-lago-di-albano/ Excursion to discover Lake Albano and its peculiarities.
Carpfishing Castelli Romani https://www.facebook.com/Carpfishing-Castelli-Romani-170702572974269/ Sport, fishing.
Federcanoa https://www.federcanoa.it/home/news/27-fick-news/5238-laghi-sicuri-il-centro-federale-di-castel-gandolfo-base-operativa-per-il-soccorso-in-acqua.html Canoe Kayak Italian Federation.
Canoa Kayak Academy https://www.ckacademy.it/ Canoa Kayak Academy.
Associazione Italiana per la lotta alle Sindromi Atassiche https://www.aisasport.it/ Canoe Sport Activities.
Latium Volcano Associazione Ecologica Ambientale https://www.latiumvolcano.it/passeggiateculturali_castelliromani.html Historical, cultural and artistic itineraries in the Castelli Romani.

Groups and associations for the protection of Lake Albano

Facebook public group: *Salviamo il Lago Albano!!!* https://www.facebook.com/groups/salviamoillagoalbano/ Most active Facebook group that offers initiatives, meetings, and proposals.

Associazione Castel Gandolfo in Movimento https://castelgandolfoinmovimento.blogspot.com/2019/07/lago-albano-di-castel-castel-gandolfo.html The purpose of the association is to raise public awareness of safety, legality, cleanliness, and respect for the public beach.
Parco dei Castelli Romani: *Associazione Bernacca Onlus* https://www.parcocastelliromani.it/inners/pages/webcam-sul-lago-albano The *Bernacca* non-profit association has activated a panoramic webcam on Lake Albano, in the municipality of Castel Gandolfo, in the Castelli Romani Regional Park.
Meteocastelli http://www.meteocastelli.it/stazioni-meteo/stazione-meteo-del-lago-albano-e-di-castelgandolfo Weather, webcam and geophysics services from Castelli Romani – Colli Albani: Weather station of Lake Albano
RESEDA S.c.s.i. http://www.resedaweb.org/salvaguai/index.htm RESEDA is a non-profit cooperative society that operates in the field of ecology, renewable energy sources, defense of the environment and for the integration of disabled people.
Associazione Amici del Parco dei Castelli Romani https://www.amiciparcocastelliromani.it/ita-chi-siamo Friends of the Castelli Romani Park Association is as an association with a naturalistic, sporting and cultural character, pursuing non-profit and social solidarity purposes.
Arco di Diana APS: *Gruppo di Studi ricerca archeologica* http://www.arcodidiana.com/ The association manages the site that hosted the Roman port on the shores of Lake Albano.
Italia Nostra: *Sezione Castelli Romani* https://www.italianostra.org/sezioni-e-consigli-regionali/lazio/castelli-romani/ The association is active in the field of the water crisis that has hit Lake Albano in the last 30 years, due to both the consumption of the soil and the increasing of resident population.
CIRF – Centro Italiano per la Riqualificazione Fluviale: *Webinar SOS Laghi Albano e di Nemi* https://www.cirf.org/it/webinar-sos-laghi-albano-e-di-nemi/ The association has organized a cycle of webinars on the Lake Albano and the Lake of Nemi that have dealt with the rapid and alarming decrease in their water level, highlighting the causes, solutions and design ideas.
AIPIN – Associazione Italiana Per l'Ingegneria Naturalistica https://www.aipin.it/#:~:text=L'AIPIN%20%C3%A8%20un'associazione,le%20tecniche%20dell'ingegneria%20naturalistica AIPIN is a non-profit technical-scientific association with cultural and professional purposes founded with the aim of disseminating the methods and techniques of naturalistic engineering.

Assonautica Acque Interne Lazio e Tevere: ***Programmazione e controllo per lo sviluppo territoriale integrato e per il turismo sostenibile*** https://assonauticalaziotevere.it/ The activity of Assonautica Acque Interne Lazio e Tevere is aimed at creating synergies with local and territorial bodies and institutions on the issues of the environment, protection, enhancement, integrated development, sustainable tourism, navigation and water sports.

Articles

Il Porto Romano sulle rive del Lago Albano torna a vivere ***(The Roman Port on the shores of Lake Albano comes back to life)*** 10 June 2022 https://www.metamagazine.it/il-porto-romano-sulle-rive-del-lago-albano-torna-a-vivere/
Castel Gandolfo, 29 maggio pulizia del Lago Albano con Carpfishing Castelli Romani ***(Castel Gandolfo, 29 May cleaning of Lake Albano with Carpfishing Castelli Romani)*** 15 May 2022 https://www.castellinotizie.it/2022/05/15/castel-gandolfo-29-maggio-pulizia-del-lago-albano-con-carpfishing-castelli-romani/
S.O.S. Lago Albano di Castel Gandolfo. Consorzio di Bonifica Litorale Nord di Roma: ***Pronti a mettere a disposizione la competenza dei nostri tecnici. (S.O.S.Lake Albano of Castel Gandolfo. North of Rome Coastal Reclamation Consortium: Ready to make available the expertise of our technicians.)*** 5 May 2022 https://www.osservatoreitalia.eu/s-o-s-lago-albano-di-castel-gandolfo-consorzio-di-bonifica-litorale-nord-di-roma-pronti-a-mettere-a-disposizione-la-competenza-dei-nostri-tecnici/
I danni causati dall'uomo all'ecosistema del Lago Albano ***(Manmade damage to the ecosystem of Lake Albano)*** 4 April 2022 https://ilgiornaledellambiente.it/danni-causati-uomo-ecosistema-lago-albano/
Filmate possibili emissioni gassose sul Lago Albano ***(Possible gas emissions filmed on Lake Albano)*** 8 December 2021 https://www.controluce.it/notizie/filmate-possibili-emissioni-gassose-sul-lago-albano/
Rafforzare le politiche di tutela su tutto il territorio lacustre dei Castelli Romani ***(The protection policies throughout the Castelli Romani lake territory have been strengthened)*** 29 July 2021 https://golettaverde.legambiente.it/2021/07/29/rafforzare-le-politiche-di-tutela-su-tutto-il-territorio-lacustre-dei-castelli-romani/
Castel Gandolfo: Lago Albano più pulito grazie ai cittadini e alle associazioni ***(Castel Gandolfo: Lake Albano cleaner thanks to citizens and associations)*** 13 April 2021 https://ilcaffe.tv/articolo/23103/lago-albano-piu-pulito-grazie-ai-cittadini

<table>
<tr><td>La crisi idrica dei laghi di Albano e Nemi – Attività di pianificazione sui Colli Albani negli ultimi vent'anni (The water crisis of the Lakes Albano and Nemi – Planning activities on the Colli Albani in the last twenty years)
April 2021 – Professione geologo: Notiziario dell'ordine dei geologi del Lazio (Professione geologo: newsletter of the order of geologists of Lazio Region)
extension://nhppiemcomgngbgdeffdgkhnkjlgpcdi/data/pdf.js/web/viewer.html?file=https%3A%2F%2Fgeologilazio.it%2Fwp-content%2Fuploads%2F2021%2F04%2FPG-62-web_new.pdf</td></tr>
<tr><td>Variazioni del livello del Lago Albano: informazioni dal fondale e ipotesi di ricostruzione storica (Changes in the level of Lake Albano: information from the seabed and hypothesis of historical reconstruction)
April 2021 – Professione geologo: Notiziario dell'ordine dei geologi del Lazio (Professione geologo: newsletter of the order of geologists of Lazio Region)
extension://nhppiemcomgngbgdeffdgkhnkjlgpcdi/data/pdf.js/web/viewer.html?file=https%3A%2F%2Fgeologilazio.it%2Fwp-content%2Fuploads%2F2021%2F04%2FPG-62-web_new.pdf</td></tr>
<tr><td>Festa della Madonna del Lago (Feast of the Madonna del Lago)
24 August 2019
https://www.castellinforma.it/home/event/92004</td></tr>
<tr><td>Lago Albano di Castel Gandolfo: Le iniziative dell'associazione Castel Gandolfo in movimento (Lake Albano of Castel Gandolfo: The initiatives of the Castel Gandolfo in motion movement)
28 July 2019
https://castelgandolfoinmovimento.blogspot.com/2019/07/lago-albano-di-castel-castel-gandolfo.html</td></tr>
<tr><td>Lago Albano osservato speciale. Arriva l'Istituto Superiore di Sanità (Lake Albano under special scrutiny. Here comes the Higher Institute of Health)
7 March 2019
https://ilcaffe.tv/articolo/52857/lago-albano-osservato-speciale</td></tr>
<tr><td>Allarme a Castelgandolfo, i fondali del Lago Albano sono pieni di bombe (Alarm in Castel Gandolfo, thedepths of Lake Albano are full of bombs)
11 November 2016
https://www.ilmattino.it/primopiano/cronaca/castelgandolfo_allarme_bombe_fondali_pieni-2074934.html?refresh_ce</td></tr>
<tr><td>Lago Albano: dentro l'antico emissario (Lake Albano: inside the ancient emissary)
Luglio/Agosto 2016 – Archeologia Viva n. 178
https://www.archeologiaviva.it/4496/lago-albano-dentro-lantico-emissario/</td></tr>
<tr><td>Albano naviga in cattive acque (Albano is in dire straits)
13 November 2012
https://comune-info.net/albano-naviga-in-cattive-acque/</td></tr>
<tr><td>Albano Laziale Emissario del lago (Albano Laziale Emissary of the lake)
https://www.tesoridellazio.it/tesori/albano-laziale-emissario-del-lago/</td></tr>
</table>

È realtà il *Contratto di Falda Lago Albano, Nemi e per il fiume Incastro.* Enti e Associazioni insieme per salvare i laghi dei Castelli Romani *(The Aquifer Contract of Lake Albano, Nemi and for the Incastro river has become reality. Bodies and associations join forces to save the Castelli Romani lakes)* https://www.castellinotizie.it/2021/03/02/e-realta-il-contratto-di-falda-lago-albano-nemi-e-per-il-fiume-incastro-enti-e-associazioni-insieme-per-salvare-i-laghi-dei-castelli-romani/
Lago Albano *(Lake Albano)* https://www.ricominciodaroma.it/item/lago-albano/

REFERENCES

[1] Patera, A. & Fabbri, A.G., Analisi della pericolosità di frana dei versanti interni del Lago Albano (Landslide hazard assessment of the internal slopes of Lake Albano). *Rapporti Tecnici INGV*, Rome (in press).

[2] Barberi, F., Chelini, F., Marinelli, G. & Martini, M., The gas cloud of Lake Nyos (Cameroon, 1986): Results of the Italian technical mission. *Journal of Volcanology and Geothermal Research*, **39**, pp. 125–134, 1989.

[3] Funiciello, R., Giordano, G., De Rita, D., Carapezza, M.L. & Barberi, F., L'attività recente del cratere del Lago Albano di Castelgandolfo. *Rendiconti dell'Accademia dei Lincei (Scienze Fisiche e Naturali)*, **9**(13), pp. 113–143, 2002.

[4] Carapezza, M.L. & Tarchini, L., Magmatic degassing of the Alban Hills volcano (Rome, Italy): Geochemical evidence from accidental gas emission from shallow pressurized aquifers. *Journal of Volcanology and Geothermal Research*, **165**, pp. 5–16, 2007.

[5] Bozzano, F., Mazzanti, P., Anzidei, M., Esposito, C., Floris, M., Bianchi Fasani, G. & Esposito, A., Slope dynamics of Lake Albano (Rome, Italy): Insights from high resolution bathymetry. *Earth Surface Processes and Landforms*, **34**, pp. 1469–1486, 2009.

[6] Mazzanti, P., Bozzano, F. & Esposito, C., Submerged landslide morphologies in the Albano Lake (Rome, Italy). *Submarine Mass Movements and Their Consequences*, eds V. Lykousis, D. Sakellariou & J. Locat, Springer: Dordrecht, 2007.

[7] Anzidei, M., Esposito, A. & De Giosa, F., The dark side of the Albano crater lake. *Annals of Geophysics*, **49**, pp. 1275–1287, 2007.

[8] Cruden, D.M. & Varnes, D.J., Landslide types and processes. *Landslides Investigation and Mitigation*, eds A.K. Turner & R.L. Shuster, National Research Council, Transportation Research Board: Washington, DC. Special Report, **247**, pp. 36–75, 1996.

[9] Mulder, T. & Cochonat, P., Classification of offshore mass movements. *Journal of Sedimentary Research*, **66**(1), pp. 43–57, 1996.

[10] Chung, C.J. & Fabbri, A.G., Representation of geoscience data for information integration. *Natural Resources Research*, **2**, pp. 122–139, 1993.

[11] Fabbri, A.G. & Chung, C.J., How credible is my hazard map? Dissecting a prediction pattern of landslide susceptibility. *WIT Transactions of Engineering Sciences*, vol. 121, WIT Press: Southampton and Boston, pp. 3–19, 2018. DOI: 10.2495/RISK180011.

[12] Fabbri, A.G. & Patera, A., Spatial uncertainty of target patterns generated by different prediction models of landslide susceptibility. *Applied Sciences*, **11**, p. 3341, 2021.

[13] Chung, C.J. & Fabbri, A.G., Prediction models for landslide hazard using fuzzy set approach. *Geomorphology and Environmental Impact Assessment*, eds M. Marchetti & V. Rivas, Balkema: Rotterdam, pp. 31–47, 2001.

[14] Chung, C.F., Computer program for the logistic model to estimate the probability of occurrence of discrete events. *Geological Survey of Canada Paper*, 78-11, pp. 1–23, 1978.

[15] Davis, J.C., Davis, J.C., Chung, C.F. & Ohlmacher, G.C., Two models for evaluating landslide hazards. *Computers and Geosciences*, **32**, pp. 1120–1127, 2006.

[16] Chung, C.J. & Fabbri, A.G., Validation of spatial prediction models for landslide hazard mapping. *Natural Hazards*, **30**, pp. 451–472, 2003.

[17] Fabbri, A.G., Cavallin, A., Patera, A., Sangalli, L. & Chung, C.J., Comparing patterns of spatial relationships for susceptibility prediction of landslide occurrences. *Advancing Culture of Living with Landslides: Advances in Landslide Science*, eds M. Mikoš, B. Tiwari, Y. Yin & K. Sassa, Springer: Switzerland, pp. 1135–1144, 2017.

GEOSTATISTICAL MODELING OF SEISMIC ACTIONS ON THE STRUCTURAL COMPONENTS OF THE SAN BENEDETTO ROAD TUNNEL, ITALY

MASSIMO GUARASCIO, ANGELO LIBERTÀ, DAVIDE BERARDI,
EMIN ALAKBARLI & MARA LOMBARDI
Sapienza University of Rome, Italy

ABSTRACT

The Italian legislation requires determining the project seismic actions and to carry out the dynamic stability verification of the structural elements of a building or road infrastructure on the base of the seismic hazard curve of the construction site. The geographic density of the available seismic data requires the best use of existing data and above all not introducing phenomenological assumptions and unverified information into the survey. This article proposes an investigation protocol that integrates the multivariate statistical analysis methodologies already used to determine the seismic shaking attenuation with more efficient and versatile geostatistical methodologies for a more realistic estimate of non-stationary parameters. The application of the new investigation protocol to the earthquakes that occurred in Italy in 2016 made it possible to detect and resolve three fundamental aspects of seismic modeling: (i) the recorded data highlight the presence of a directional anisotropy of the arrival time and the value of the acceleration peak on the ground which led to the introduction of a non-Euclidean metric in seismic modeling; (ii) the presence and measurement of the geographical continuity of the irregular variations between data pairs attributable to the heterogeneity of the rock formations and outcropping soils; (iii) the need for an experimental measure of the estimate uncertainty for an objective evaluation of the applied numerical estimator. This last result made it possible to evaluate the gain in terms of accuracy of the estimate performed with the local geostatistical estimator. The article presents the estimate of the arrival time and the peak value of the vertical (PGA_V) and maximum horizontal (PGA_H) component of the acceleration of the vibrational movements of the rock around the San Benedetto tunnel during the Norcia earthquake. The maximum estimated values of PGA_V and PGA_H are around the section of the tunnel damaged by the earthquake.

Keywords: earthquake, geostatistics, Universal CoKriging, seismic first-guess, risk, analysis, road tunnel, seismicity.

1 INTRODUCTION

The Italian legislation (NTC 2018) limits the design of buildings and road infrastructures (viaducts, bridges, tunnels, etc.) to the seismic hazard of the construction site. The seismic actions of the project must be determined on the hazard curve by identifying the accelerations on the ground whose return times or the probabilities of exceeding during the life of the building are prescribed by the legislation and referred to the operating limit states, life protection and collapse of the building or infrastructure. In most cases, the centenary historical series of earthquakes that occurred in the seismogenetic area and the seismic monitoring data recorded during the most recent earthquakes are the only information available to plot the probability curves of the accelerations perceived by the building or infrastructure for future earthquakes in the geographic area of the construction site.

The limited number of data makes to use efficient objective methods of analysis and to not introduce into the seismic investigation phenomenological assumptions and unverified information. To this end, the authors propose an investigation protocol to perform an objective and verifiable estimate of the non-stationary ground-motion parameters and to draw the anisotropic ground-motion attenuation diagrams using at best the data recorded by a

WIT Transactions on The Built Environment, Vol 214, © 2022 WIT Press
www.witpress.com, ISSN 1743-3509 (on-line)
doi:10.2495/SSR220021

seismic monitoring network. Lithological data of rock formations and soils outcropping can be used if georeferenced and homogeneously detected. Below are the six phases of the seismic investigation:

(i) identify data recorded from the accelerometer stations the local main directions of seismic oscillations during earthquake and the local maximum ground shaking intensity (multivariate-statistical analysis);
(ii) identify the function of the geographic seismic patterns consistent with the theory of elastic wave propagation and the georeferenced seismic available data (first-guess function);
(iii) determine the geographic continuity model of the ground motion parameters to be estimated (statistical inference of the variogram function);
(iv) estimate the ground motion parameters in the station points using the best available data and the topographic model of the investigation area (cross-validation of no stationary multivariate geographical estimator);
(v) use the validated estimator to estimate the ground motion parameters in any geographical point using the best available data and the topographic model of the investigation area;
(vi) starting from the estimated seismic data at the nodes of a georeferenced grid, mark the directional attenuation profiles of the ground motion parameters of interest.

The above investigation protocol was applied to estimate, at the nodes of a 500 × 500 m^2 georeferenced grid, the arrival time and the value of the peak of the horizontal (PGA_H) and vertical (PGA_V) component of ground acceleration of the four earthquakes that occurred in 2016 in the central Italy: Accumoli (24 August 2016), Amatrice (26 August 2016), Castelsantangelo sul Nera (26 October 2016) and Norcia (30 October 2016).

The investigations carried out on the four earthquakes revealed three fundamental results not found in the scientific literature of the sector. Around the area of the maximum ground motions (epicentral area) the geographic pattern of the ground motion parameters is approximated by a continuous function (first-guess seismic) which is defined by the product of the geometric attenuation equation of an elastic wave emitted by a point source and the inelastic attenuation equation [1]. For each earthquake of 2016, the optimal adjustment of the *FgS* function obtained by minimizing the squares of the seismic residuals calculated as the difference between the PGA at the station points and the values of the *FgS* function at the same points, presents a directional anisotropy. This led to the modify the shape of the geometric attenuation equation and of the inelastic attenuation equation, introducing a non-Euclidean spatial metric for the first time in seismic modeling.

The second result is the measurement of the geographical continuity of the two seismic variables treated (arrival time and value of the PGA). This geographical property is likely inherited from the continuity of the lithological characteristics of the rock formations and surface soils crossed by the volume seismic waves (waves P and S) and by the surface seismic waves. More or less large volumes of rock with homogeneous or slightly variable lithological characteristics are in contact with volumes of rock with greater spatial variability or with different lithological characteristics. The average extension of the lithological discontinuities determines the geographical variability of the data recorded by the accelerometric stations.

This leads to the rejection of the hypothesis of random distribution of local seismic variations and, although locally the seismic variations are irregular, the average of the squared differences of the ground motion parameters recorded in two geographical points tends to zero when the distance between the points tends to zero.

The correlation between lithological characteristics and PGA probably will be found in other ground motion parameters such as the spectral parameters of the accelerograms (SA) and the seismic indicators of engineering interest: Arias Intensity Index and Cosenza–Manfredi Index [2], which can then be analyzed with the proposed investigation protocol.

The identified *FgS* functions are compatible with the theory of propagation of inelastic waves within a rock model whose lithological properties are likely a spatial average of the lithological properties of the rock volumes crossed by the seismic waves. The more the rock formations are heterogeneous, the higher is the geographic variability of the residues of the FgS function. The geographic variability of seismic residues is proportional to the lithological heterogeneity of the volumes of rock crossed by the seismic waves. For the authors, the variance of local dispersion of seismic residues is a correct indicator of the complex and undetectable geometry of the stratigraphic surfaces and the heterogeneity of rocks and soils outcropping [3]. And the origin of the local indeterminacy of ground shaking intensity is the unbridgeable difference between the amount of information needed to solve the complexity of the geostratigraphic system with sufficient approximation and the seismic information contained in the available data. The local indeterminacy of the estimate is quantified by the variance of the estimation error, which is calculated by the same no-stationary geostatistical estimator (Universal CoKriging (UCoK) with external drift). For each data detected around the point to be estimated, the UCoK calculates a weight coefficient that depends on: (i) the local function of geographical continuity of the seismic variable; (ii) geometry of the seismic monitoring network; (iii) position of the accelerometric stations with respect to the point to be estimated. Therefore the local indeterminacy of the estimate does not depend only on the distance of the point to be estimated from the epicentral area, as predicted by the logarithmic estimator proposed by Campbell [1] and subsequently applied by Boore et al. [4] and by Sabetta and Pugliese [5] but, as aforementioned, also from the number and position of the data collected around each point to be estimated.

The third result that emerged from the survey is the measurement of the gain in terms of accuracy of the local estimate obtained with the geostatistical estimator compared to the estimate obtained with the *FgS* function. The improvement in the estimate is measured by the difference between the variance of the seismic residuals from the *FgS* function in the station points and the variance of the local estimation error in the same station points. In phase IV of the seismic survey, the UCoK estimator is used to calculate the ground-motion parameters in the station points, eliminating the station data to be estimated from the calculation each time. In each station point and for each ground motion parameters are therefore available: (i) the data detected (data known only in the station points); (ii) the seismic residue of the datum detected by the *FgS* function (iii) the datum estimated with the UCoK estimator; (iv) the estimation error committed by the UCoK (difference between true data and estimated data; (v) variances of the estimation error calculated by the same UCoK estimator. For each of the four Italian earthquakes it was found that:

(i) the *FgS* function and the UCoK are both unbiased estimators (the average of the seismic residues and the average of the estimation errors are close to zero);
(ii) the variance of dispersion of the estimation errors committed with the UCoK was always lower than the variance of dispersion of the seismic residuals referred to the function of *FgS*;
(iii) the variance of the estimation error calculated by the UCoK is very close to the variance of dispersion calculated on the real estimation errors.

It follows that the function of *FgS* is a first approximation of the attenuation of ground motions and cannot be assumed a priori as the prediction equations of ground shaking intensity as proposed by Campbell [6], Boore et al. [4] and Sabetta and Pugliese [5].

The paper show how estimate the arrival time and the peak value of the vertical (PGA_V) and horizontal (PGA_H) component of the acceleration of the rocks around the San Benedetto tunnel during the Norcia earthquake (30 October 2016)

2 GEOSTATISTICAL FRAMEWORK

Initially, geostatistics was developed to estimate mineral deposits [7] but the same methodologies were found to be valid and therefore currently applied in many fields of earth sciences (hydrogeology, meteorology, oceanography, geochemistry, environmental control, soil science, etc.). The basis of geostatistical methodologies is the observation that some natural phenomena have a geographical evolution and local states are closely related to their position. This typology of phenomena has been defined as "regionalized phenomena" and consequently the variables that describe them are defined as "regionalized variables" (V.R.) [8].

The states assumed by a V.R. at points x of a geographic domain are represented by a function *z(x)*. Locally, the *z(x)* function is highly irregular and the variations from one point to another are often unpredictable, while over greater distances the function is more continuous and shows the structural characteristics of the physical phenomenon that underlies the observed regionalized phenomenon. This feature suggested developing the function *z(x)* into a sum of two functions; the first, indicated by *m(x),* which represents the geographical trend of the regionalized phenomenon and the second, indicated by *y(x),* which reflects the irregular local variations of the observed phenomenon: $z(x) = m(x) + y(x)$. As they are defined, the two components are statistically independent.

This double aspect is found in numerous geophysical and geochemical variables such as: temperature of the layer of atmosphere in contact with the earth's surface (temperature decreases with latitude and topographical altitude), concentration of pollutants in the soil (concentration increases near an oil plant) or vibrations of the ground following a seismic event (vibrations of the ground decrease with the distance from the epicentral area of the earthquake).

The irregular geographical evolution of natural phenomena has suggested the conceptual choice of considering the R.V. z(x) of the realizations of Random Functions (R.F.) Z(x) defined on a precise geographical domain D of the Euclidean space $\mathbb{R}^n$. The function m(x) represents at each point x the expected value of the R.F. Z(x): $E|Z(x)|=m(x)$. The residual Y(x) of the R.F. Z(x) is a locally stationary R.A. with $E|Y(x)|=0$ and covariance $Cov|Y(x_1), Y(x_2)| = E|Y(x_1)\, Y(x_2)|<\infty$. In the study of the natural phenomena described by several V.R. related, information on the geographical structure of the R.F. $Z_i(x)$ e $Z_j(x)$ is developed by the variograms and cross-variograms of the R.F. of the residuals $Y_i(x)$ and $Y_j(x)$:

$$\gamma_{i,j}(\vec{h}) = \frac{1}{2}\boldsymbol{E}|[(Y_i(x+\vec{h}) - Y_i(x)][(Y_j(x+\vec{h}) - Y_j(x)]|. \tag{1}$$

Universal CoKriging with external drift is a specialization of Universal CoKriging (UCoK) applied when the functions that describe the geographic pattern of the regionalized variables studied (external drift) are known. UCoK is widely used in earth sciences to map geophysical or geochemical variables such as the minimum and maximum air temperature or the concentration of metals or pollutants in the soil. The UCoK estimator is numerically equivalent to a multivariate geographic regression conditional on the directional variogram and cross-variogram functions of the target variable and the auxiliary variables. The estimate of the target variable Z_o at point x is obtained from the linear combination (weighted average) of the data of the target variable Z_o and the auxiliary variables Z_i which have a geographical correlation with the target variable.

The universal CoKriging with external drift assumes that (i) the geographic patterns of the R.F. objective and the R.F. auxiliaries are in every point of the geographic domain equal to $E|Z_i(x)| = \sum_{l=0}^{L} a_{i,l}\, f_i^l(x)$; (ii) for each pair of points in the geographic domain, the mean of the squared differences of the R.F. and the average of the product of the difference between two R.F. are determined by the variogram and cross-variogram model.

For each point to be estimated the estimation weigh coefficients of the data $\lambda_{k,i}$ are obtained by setting the following conditions:

(i) average estimation error of the zero target variable: $\boldsymbol{E}|[(Z_0^*(x) - Z_0(x)]| = 0$
The condition of unbiased for the UCoK estimator becomes: $\boldsymbol{\lambda}_i^T\ \boldsymbol{F}_{i,l}\ = \delta_{i,l}\,\boldsymbol{f}_i^T$

(ii) variance estimator error of the minimum target variable:

$$\boldsymbol{Var}|[(Z_0^*(x) - Z_0(x)]| = 2\ \textstyle\sum_i \boldsymbol{\lambda}_i^T\ \boldsymbol{\Gamma}_{i,0} - \sum_i \sum_j \boldsymbol{\lambda}_i^T\ \boldsymbol{\Gamma}_{i,j}\ \boldsymbol{\lambda}_j\ = \text{min.}$$

3 NORCIA EARTHQUAKE OF 30 OCTOBER 2016

This section presents the results of the seismic survey of the Norcia earthquake of 30 November 2016 (seismogenic zone of the Central Italian Apennines) performed to determine the shaking intensity that the San Benedetto tunnel underwent during the seismic event. The UCoK estimator was used to estimate the seismic variables time of arrival and peak value of the vertical (PGA_V) and maximum horizontal (PGA_H) component of the acceleration of the vibrational movements of the rock. The two seismic variables were estimated at the nodes of a three-dimensional grid with a distance between the nodes of 13.5 m in the direction of the longitudinal axis of the tunnel and 27.0 m in the transverse plane. The extension of the grid is 4,117.5 m in the direction of the longitudinal axis of the tunnel and 216 × 216 m^2 in the transverse plane.

The basic data used to estimate the 2 seismic variables are: (i) the seismograms recorded by the accelerometric stations; (ii) the geographic coordinates and the altitude of the accelerometric stations; (iii) the geo-referenced altimetric model with 500 m cells (DTM); (iv) the geographical coordinates and the height of the grid nodes around the San Benedetto tunnel.

3.1 Local dimensional ground motions seismic.

The effects and the failure of the structural elements of a building caused by seismic motions depend both on the intensity of the shaking and on the directions of oscillation of the ground during an earthquake. For the purpose of verifying the dynamic stability of the structural elements of a building or road infrastructure it is useful to know the directions of the principal axes of a 3D-orthogonal system along which the component of ground motion have maximum, minimum and intermediate values of variance and have zero values of covariance. The methodology proposed by Pazien and Watabe [9] was used to determine, starting from the accelerograms recorded by the stations of the seismic monitoring network of Central Italy, the local directions of the three main axes of ground motions during the Norcia earthquake. The method used usually to compute the principal axes is by diagonalized the covariance matrix of ground accelerograms recorded along the instrument axes:

$$\boldsymbol{C}(t) = \begin{vmatrix} c_{x,x} & c_{x,y} & c_{x,z} \\ c_{y,x} & c_{y,y} & c_{y,z} \\ c_{z,x} & c_{z,y} & c_{z,z} \end{vmatrix} \tag{2}$$

where $c_{i,j}(t) = E|a_i(t)\, a_j(t)|$ and $a_i(t)$ are random functions which represented accelerograms recorded along east–west, north–south and vertical direction. The time interval is 0.005 seconds.

The directions of the principal axes are the eigenvectors computed through the use of the covariance matrix $C(t)$ and the principal variances are the corresponding eigenvalues. It has been verified that the time-correlation of the accelerations $a_i(t)$ is zero after 0.01 seconds and in the same time-intervals the average of the accelerations is close to zero.

Since ground motions are non-stationary over time a moving window technique is applied to analyze the time-dependent characteristic of the direction of principal axes. The eigenvalues and eigenvectors are calculated on disjoint intervals of 0.25 seconds using the 200 data recorded in the interval of one second centered in the target interval.

In the fifteen accelerometric stations closest to the epicentral area of the Norcia earthquake (2016), the main axis with the least variance is sub-vertical in the interval time of maximum seismic shaking (Fig. 1). In the same interval the other two principal axes are naturally sub-horizontal. For these two axes the direction is not always constant over time. For the axes of major and intermediate values of variance, a statistically prevalent direction was not found. Although for a few stations, such as that of Forche Canapina (identified by the FCC code), a prevailing direction can be identified (Fig. 1).

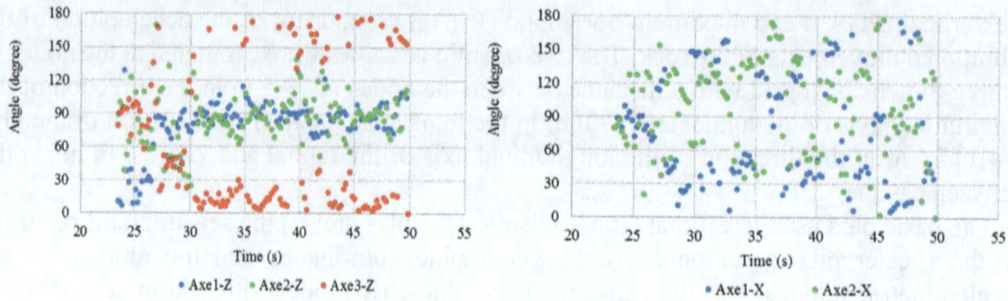

Figure 1: Scatter diagrams: Angle between principal axes and the two direction vertical and east–west (FCC station).

The scatter-diagram of the first graph in Fig. 1 shows the angles that the main axes form with the vertical direction of the FCC station. The direction of the axe with minimum value of variance is parallel to the vertical direction (red dots) while the other two main axes are orthogonal to the vertical direction. The second graph in Fig. 1 shows the angles that the axes with maximum and intermediate values of variance form with the east–west direction. The prevailing direction of the main axes 1 and 2 close to 45° and 135° respectively can be seen from the scatter diagram of second graph in Fig. 1.

3.2 First-guess of seismic variables

The first-guess fields of the two seismic variables are described by continuous mathematical functions defined at all points of the geographical survey domain.

In geostratigraphic systems, the regionalization of the ground motion parameters differs from the geographical pattern predicted from the theory of propagation of anelastic waves in a homogeneous material due to the heterogeneity of the rock formations crossed and the complex geometry of the fault systems and the stratigraphic contacts between lithological units:

- the different propagation velocity of seismic waves in rocks deform the geometry of the wave surfaces with consequent directional variations of the geometrical attenuation;
- moreover, irregular local variations of the seismic oscillations are a consequence of the different anelastic behavior of the rocks. A part of the energy carried by the seismic waves is continuously and irregularly dissipated by friction (anelastic attenuation).

Despite this, the data of the seismograms recorded by the accelerometric stations show good consistency with the PGA geographic pattern (Fig. 2) and the arrival time of the acceleration peak (Fig. 3) described by the functions of *FgS*.

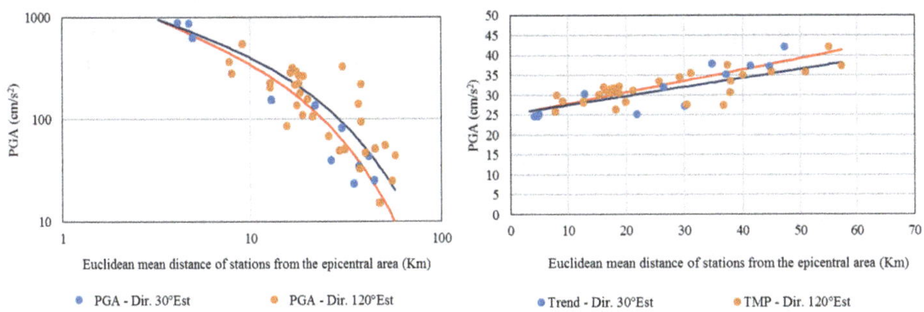

Figure 2: First-guess of PGA_V and arrival time in Euclidean distance.

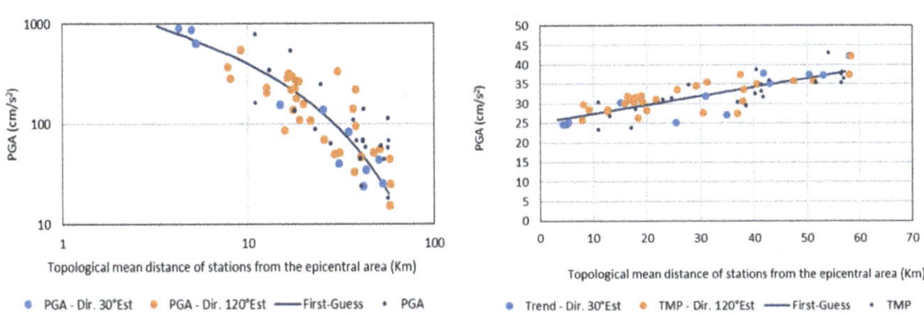

Figure 3: First-guess of PGA_V and arrival time in topological distance.

The scatter-diagrams of the arrival time and the value of the acceleration peak on the ground recorded by the stations in the direction orthogonal to the Apennine axis (probable fault direction) (red points) are mainly arranged in the lower part (PGA_V) and in the high (arrival time) of the cloud. In the graphs, the abscissa axe indicates the Euclidean mean distance from the epicentral area. This directional anisotropy also emerged in the numerical simulations performed to estimate the coefficients of the *FgS* function.

First-guess function of the arrival time of the acceleration peak

$$f_t(x) = a_t(x_0) - b_t(x_0)\,\overline{r}_0(x), \tag{3}$$

where x_0 identifies the pair of coordinates of the center of gravity of the epicentral area of the seismic event; $a_t(x_0)$ and $b_t(x_0)$ are the coefficients of the regression line; $\overline{r}_0(x)$ is the average topological distance of the x coordinate point from the epicentral area (average of the topological distances between the point and the points of the circle with a radius of 5 km centered at the point x_0) [8].

First-guess function of the peak ground acceleration (PGA)

$$f_p(x) = \frac{a_p(x_0)}{\sqrt{\overline{r}_0(x)}}\, e^{-b_p(x_0)\,\overline{r}_0(x)} \tag{4}$$

where: x_0 identifies the pair of coordinates of the center of gravity of the epicentral area of the seismic event; $a_p(x_0)$ and $b_p(x_0)$ are the coefficients of the FgS function, which it describes the geographical pattern of the PGA; $\overline{r}_0(x)$ is, as for the seismic variable time of arrival, the average topological distance of the point of coordinates x from the epicentral area.

In the two first-guess functions the topological distance between any pair of points x' and x" is defined as: $d(x',x'') = \sqrt[2]{\boldsymbol{X}^T \boldsymbol{H}\, \boldsymbol{X}}$ where $\boldsymbol{X} = (x_1'' - x_1', x_2'' - x_2')^T$ is the difference vector of the metric coordinates of the two points, and $\boldsymbol{H} = \boldsymbol{\psi}\, \boldsymbol{M}\, \boldsymbol{\psi}^T$ is the product of the rotation matrix of the cartesian reference system and of the metric tensor M used to introduce in the first-guess functions the directional anisotropy of the seismic variables:

$$\boldsymbol{M} = \begin{vmatrix} (1+\alpha)^2 & 0 \\ 0 & 1 \end{vmatrix}. \tag{5}$$

The coefficient α is the expansion factor of the metric coordinates of the Cartesian axis orthogonal to the Apennine alignment (oriented at 35° east).

The position of the epicentral area must be compatible with the geographic pattern described by the $m_p(x)$ function and with the PGA data at the points of the accelerometric stations.

To calculate the coefficients of the *FgS* function of the PGA, a Monte Carlo simulation is performed, which consists of N random extractions of the inelastic damping coefficient, of the angle between the direction of greatest geometric attenuation and the East direction and of the coefficient a of the metric tensor M for each node of the 500 × 500 m^2 grid. The numerical values extracted are compatible with the seismic phenomenon to be modeled. For each extraction, the coefficient ap is determined (adjustment factor to the existing seismic data) which minimizes the difference between the PGA data at the station points and the following function (multivariate geographic regression):

$$m_p(x) = c_0 + c_q q(x) + \frac{a_p(x_0)}{\sqrt{\overline{r}_0(x)}}\, e^{-b_p(x_0)\,\overline{r}_0(x)} \tag{6}$$

where x_0 is the coordinate of the grid node and *q(x)* is the topographic elevation at the *x* coordinate point. The coefficient a_p is calculated as the weighted average of the PGA data around the node to be estimated $a_p(x_0) = \sum_{i=1}^{n} \lambda_{0,i}\, z(x_i)$ and the weight coefficients $\lambda_{0,i}$ are the solutions of the following linear system:

$$\begin{cases} \boldsymbol{I}\, \boldsymbol{\lambda}_l + \boldsymbol{F}\, \boldsymbol{\mu} & = 0 \\ \boldsymbol{F}^T\, \boldsymbol{\lambda}_l & = \delta_{k,l} \end{cases} \tag{7}$$

where $\boldsymbol{I}$ is the identity matrices *nxn*; $\boldsymbol{F}$ is the *nx3* matrix of the values of the development terms of $m_p(x)$ in the n observation points; $\boldsymbol{\mu}$ is the vector of the three Lagrange multipliers and $\delta_{k,l}$ is the operator to set equal to 1 the *k*-th term of the coefficient to be estimated.

In the scatter diagrams the arrangement of the points representing the arrival time and the value of the ground acceleration peak recorded by the 65 seismic stations is consistent with the inelastic wave propagation theory represented by the *FgS* function (continuous curve). In the graphs, the abscissa axe indicates the topological average distance from the epicentral area $d(x',x'') = \sqrt[2]{\boldsymbol{X}^T \boldsymbol{H}\,\boldsymbol{X}}$.

3.3 Estimation by Universal CoKriging

The validated UCoK estimator was used to estimate the arrival time and the peak value of the vertical and horizontal components of the ground acceleration in the area of San Benedetto tunnel. The estimation was performed using data of ten accelerometric station near the tunnel (Fig. 4) and first-guess functions of both seismic variables, variogram and cross-variogram models and digital terrain model (DTM).

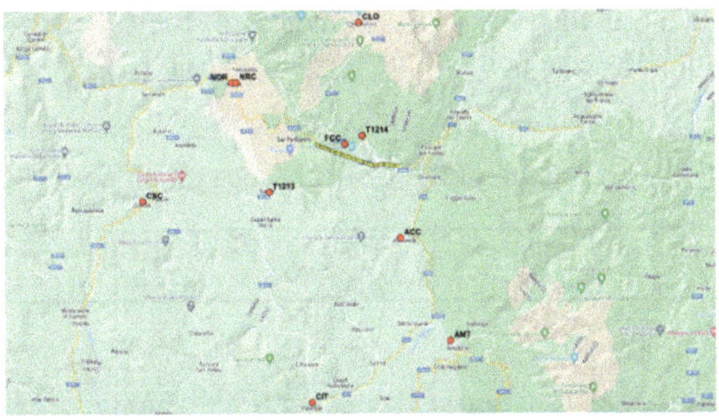

Figure 4: Position of the ten accelerometric stations used to estimate the two seismic variables at the nodes of the San Benedetto tunnel grid (in yellow the tunnel).

The estimate of the target variable Z_0 at grid-node x is obtained from the linear combination (weighted average) of the data of the target variable Z_0 and the auxiliary variable Z_i which have a geographical correlation with the target variable:

$$Z_0^*(x) = \sum_{k=0}^{m} \sum_{i=1}^{n} \lambda_{k,i}\, Z_k(x_i), \tag{8}$$

where $\lambda_{0,i}$ are the weight coefficients of the target variable and $\lambda_{k,i}$ are the weight coefficients of the auxiliary variable (index k>0) and x_j are the coordinates of the n observation points (x_j represents the pair of metric coordinates necessary to locate the *i*-th point in the geographic domain).

For each node of grid to be estimated the coefficients $\lambda_{k,i}$ are obtained from the following linear system (9):

$$\begin{cases} \sum_{k=0}^{m} -\boldsymbol{\Gamma}_{i,j}\, \boldsymbol{\lambda}_j + \boldsymbol{F}_i\, \boldsymbol{\mu}_i = -\boldsymbol{\Gamma}_{i,0} & i = 0 \dots m \\ \boldsymbol{F}_i^T\, \boldsymbol{\lambda}_i = \boldsymbol{f}_{l,0}\, \delta_{i,l} & i = 0 \dots m, \end{cases} \tag{9}$$

where $\boldsymbol{\Gamma}_{i,j}$ are the *nxn* matrices of variograms or cross-variograms of F.A. Z_i and $Z_j$between each observation points pair; $\boldsymbol{F}_{i,l}$ are the nxL matrix of the values of the first-guess functions at the observation points; $\boldsymbol{\Gamma}_{i,0}$ are the vectors of the variograms and cross-variograms between the point to be estimated (target-point) and the observation points of the F.A. to be estimated with the F.A. Z_i and $\boldsymbol{f}_{l,0}\, \delta_{i,l}$ are the vectors of the values of the first-guess function of the V.R. to be estimated and sold set to zero the conditions of unbiased of the V.R. auxiliaries.

The variance of the estimation error or cokriging variance is given by:

$$\sigma_{CK}^2 = \sum_{k=0}^{m} \boldsymbol{\lambda}_k^T \boldsymbol{\Gamma}_{k,0} + \boldsymbol{\mu}_i^T \boldsymbol{f}_{i,0}, \tag{10}$$

where $\boldsymbol{\lambda}_{i,0}$ are the vectors of the F.A. Z_i and $\boldsymbol{\mu}_{i,0}$ are the vectors of the Lagrange parameters associated with the F.A. Z_i. The variance of the estimate quantifies the uncertainty of the estimate and, as defined, depends: (i) on the number of observation points; (ii) on the geometry of the observation points; (iii) on the position of the observation points with respect to the point to be estimated and (iv) on the local variability of the V.R. treated.

The value and variance of the two seismic variables were estimated in each node of grid. In Table 1 the target variable is the PGA and the estimated value is the standard deviation of the estimation error and the relative estimation error. Table 2 shows the results of the estimation for the target arrival time variable. The two tables lists the ten accelerometric stations and for each one the distance from the estimated node and the weight coefficients of the target variable and the auxiliary variable are indicated. Tables 3 and 4 show the estimation of PGA and arrival time in some point of the tunnel.

Table 1: Estimation of the vertical acceleration peak (PGA).

Prg	ID station	Altitude (m)	Distance (Km)	PGAv (cm/s²)	Arrival time (s)	CoKriging Weight PGA	CoKriging Weight Arrival time
1	FCC	1553	1.267	893.50	24.74	1.4055	-14.3869
2	T1214	1490	2.649	632.91	25.18	0.0207	22.0470
3	CLO	1456	8.978	782.02	23.33	-0.0651	-0.4670
4	NRC	616	8.130	367.53	25.89	-0.1470	-2.8061
5	NOR	662	8.400	283.36	29.87	-0.0626	-4.2986
6	CSC	683	14.078	155.84	30.32	0.0137	-0.7633
7	T1213	860	5.214	868.89	24.67	-0.0219	1.0059
8	CIT	873	17.512	135.27	28.62	0.0409	2.2989
9	ACC	922	8.062	546.90	28.42	-0.1969	-2.6447
10	AMT	950	15.899	317.82	31.17	0.0127	0.0148
	Distance from epecentral area		**1.625**			**1.0000**	**0.0000**
			Estimated PGAv	**1011.37**			
			DevStandard error	**145.88**			
			Relative error (%)	**14.42**			

Table 2: Estimation of the arrival time of the vertical acceleration peak.

Prg	ID station	Altitude (m)	Distance (Km)	PGAv (cm/s²)	Arrival time (s)	CoKriging Weight PGA	CoKriging Weight Arrival time
1	FCC	1553	1.267	893.50	24.74	-0.0006	0.8439
2	T1214	1490	2.649	632.91	25.18	0.0016	0.1690
3	CLO	1456	8.978	782.02	23.33	0.0001	-0.0493
4	NRC	616	8.130	367.53	25.89	-0.0005	-0.0444
5	NOR	662	8.400	283.36	29.87	-0.0004	-0.0316
6	CSC	683	14.078	155.84	30.32	0.0000	-0.0291
7	T1213	860	5.214	868.89	24.67	-0.0005	0.2067
8	CIT	873	17.512	135.27	28.62	0.0002	0.0210
9	ACC	922	8.062	546.90	28.42	0.0001	-0.0847
10	AMT	950	15.899	317.82	31.17	0.0000	-0.0015
	Distance from epecentral area		**1.625**			**0.0000**	**1.0000**
				Estimated arrival time of PGA	**24.17**		
				DevStandard error	**1.61**		
				Relative error (%)	**6.66**		

Table 3: Estimation of the vertical acceleration peak (PGA).

Estimated point	Epicentral distance (Km)	Estimate PGA (cm/s^2)	DevStandard error (cm/s^2)	Relative error (%)
West portal area (San Pellegrino di Norcia)	1.679	1013.12	147.06	14.52
Tunnel lining segment at point 906,5 m	1.622	1011.69	146.22	14.45
Tunnel lining segment at point 920 m	1.625	1011.37	145.88	14.42
Tunnel lining segment at point 933,5 m	1.629	1010.97	145.45	14.39
East portal area (Capodacqua)	3.886	776.01	105.52	13.60

Table 4: Estimation of the arrival time of the vertical acceleration peak.

Estimated point	Epicentral distance (Km)	Estimate arrival time of PGA (s)	DevStandard error (s)	Relative error (%)
West portal area (San Pellegrino di Norcia)	1.679	24.05	1.69	7.03
Tunnel lining segment at point 906,5 m	1.622	24.17	1.62	6.70
Tunnel lining segment at point 920 m	1.625	24.17	1.61	6.66
Tunnel lining segment at point 933,5 m	1.629	24.17	1.61	6.66
East portal area (Capodacqua)	3.886	25.78	1.42	5.51

Fig. 5 shows the growth of the vertical acceleration peak from east to west with a maximum of 1,027.13 cm/s^2 at the level of the tunnel floor. In correspondence with the breaking segment of the road surface, the vertical acceleration peak was 1,013.17 cm/s^2.

Figure 5: Geostatistical estimate of the vertical acceleration peak in the nodes of the 3D grid 13.5 m in the direction of the longitudinal axis of the San Benedetto tunnel (yellow dots) and 27.0 m in the transverse vertical plane.

4 CONCLUSIONS

The results obtained on the four earthquakes of 2016 show how the lithological heterogeneity and the succession of rock formations and outcropping soils determined the geographical directional attenuation of the peaks of the horizontal and vertical components of the ground acceleration. This correlation between lithological characteristics and PGA will likely be found in other ground motion parameters such as the spectral parameters of the accelerograms (SA) and the seismic indicators of engineering interest.

REFERENCES

[1] Campbell, K.W., Strong motion attenuation relations: A ten-year perspective. *Earthquake Spectra*, **1**(4), pp. 759–804, 1985. DOI: 10.1193/1.1585292.

[2] Cosenza, E. & Manfredi, G., Damage indices and damage measures. *Progress in Structural Engineering and Materials*, **2**, pp. 50–59, 2000.

[3] Guarascio, M., Libertà, A., Berardi, D., Di Benedetto, E. & Lombardi, M., Seismic geostatistics application to the territorial infrastructure resilience and sustainability. *Transactions on The Built Environment*, vol. 206, WIT Press: Southampton and Boston, 2021.

[4] Boore, D.M., Joyner, W.B. & Fumal, T.E., Estimation of response spectra and peak accelerations from western North American earthquakes: An interim report, Part 2, U.S. Geol. Surv. Open-File Rep. 94-127, 40 pp, 1994.

[5] Sabetta, F. & Pugliese, A., Estimation of response spectra and simulation of nonstationary earthquake ground motions. *Bulletin of the Seismological Society of America*, **86**(2), pp. 337–352, 1996. DOI: 10.1785/BSSA0860020337.

[6] Campbell, K.W., Strong motion attenuation relations. *International Handbook of Earthquake and Engineering Seismology, Part B*, pp. 1003–1012, 2003.

[7] Guarascio, M., Huybrechts, C.J. & David, M., Advanced Geostatistics in the Mining Industry. *Proceedings of the NATO Advanced Study Institute*, 13–25 Oct., Istituto di Geologia Applicata of the University of Rome, Italy, 1975.

[8] Matheron, G., Random sets theory and its applications to stereology. *Journal of Microscopy*, **95**, pp. 15–23, 1972. DOI: 10.1111/j.1365-2818.1972.tb03708.x.

[9] Pazien, J. & Watabe, M., Characteristics of 3-dimensional earthquake ground motions. *Earthqu. Eng. Struct. Dyn.*, **3**, pp. 365–373, 1975.

ROAD TUNNEL RISK-BASED SAFETY DESIGN METHODOLOGY BY GU@LARP QUANTUM RISK MODEL

MASSIMO GUARASCIO, DAVIDE BERARDI, CARLOTA DESPABELADERA, EMIN ALAKBARLI, ELEONORA DI BENEDETTO, MARTA GALUPPI & MARA LOMBARDI
Sapienza University of Rome, Italy

ABSTRACT

The ALARP concept is used in different countries for different sectors of activity where a risk assessment or measure is requested. In this paper a model is developed based upon ALARP principle for tunnel risk-based design in case of fire accident scenarios. In Italy, ALARP risk acceptability and tolerability criteria have been adopted then the compliance with them has to be verified in order to guarantee a minimum-sufficient level of safety. The quantum of risk coupled with any design scenario is defined and modelled and the consequent individual quantum of risk coupled with the single exposed unit in the scenario is defined too. The methodologies for the identification of the requested design scenario, in number and type, are outlined. The scenarios are described in a shape suitable as INPUTS in the thermo-dynamical numerical simulations for fire generation and exposed units evacuation. The expected OUTPUTS of the numerical simulation are the estimations of the number of the fatalities (N) coupled with the single specific scenarios. In parallel with the above physical deterministic scenario simulations, a conceptual and operational procedure has been also established for the scenarios probabilities assessment. Merging the resulting data of both the above separate models, the risk quanta Gu@larp model is finally established. A case study is developed considering scenarios related to a virtual limit tunnel to support the description of the model itself, properties, advantages and perspectives.

Keywords: Gu@larp, ALARP, risk quanta, risk-based design, risk acceptability, risk tolerability, risk line, fire-accident rate, expected fatalities, road tunnel safety.

1 INTRODUCTION

In this paper we developed a model based upon the ALARP principle for risk-based design for fire accident scenario in a tunnel compliant with the acceptability and tolerability criteria adopted in Italy in order to guarantee a minimum-sufficient level of safety. The methodologies for the identification of the requested design scenario, in number and type, are outlined. The major key factors for the scenario identification are enumerated as follows: (1) fire rate and fire design; (2) temporal and seasonal characteristics of the traffic; (3) location of the fire and people inside the tunnel; and (4) availability of performances of the protection systems.

Each one of the identified scenarios (S_i) is also coupled with the estimation of its probability of occurrence ($P(S_i)$) estimated with the parallel established procedure based on the tree diagram.

Each deterministic calculation ($N(S_i)$=number of fatalities in the S_i) provided by FDS simulator is our estimated value of the number of fatalities compared with the number of exposed units that are located within the distance of influence (potential damages) of the fire accident.

Merging the result data of both the above separate models, the risk quanta Gu@larp model [1], [2] is finally established.

WIT Transactions on The Built Environment, Vol 214, © 2022 WIT Press
www.witpress.com, ISSN 1743-3509 (on-line)
doi:10.2495/SSR220031

A case study is developed considering scenarios related to a virtual limit tunnel [2]–[5] to support the description of the model itself, the properties and advantages, and the perspectives.

2 SCENARIO IDENTIFICATION

Quantitative risk analysis has been carried out to identify and measure the quantum of risk of a unidirectional virtual tunnel having a length of 501 m and AADT of 7,500 vehicles per lane per day (two lanes). The focus of this paper are the risks created by fires occurring inside the tunnel that have the potential of causing injuries or fatalities to the persons inside the tunnel. The fire hazards and their consequences, as the number of fatalities ($N(S_i)$), are quantified for checking the requested compliance based on the estimation of the quantum of risk measure and not on the number of fatalities that is an unknown and not available information for direct estimation. In order to do this, in appropriate number of scenario types have to be identified and probabilized through the combination of the occurrence rate of a fire in a tunnel, the length of the tunnel, the traffic density, and finally the key factors that can influence the measure of the quantum of risk of a given scenario. But since an infinite number of scenarios can be derived from the combination of different conditions and factors, it is necessary to identify and select a limited number of scenarios that can be allowed to appropriately represent all possible scenarios. Once the fixed characteristic of the tunnel such as geometry (shape, slope, etc.) and structural materials are known, the major key factors in scenarios identification are:

1. Fire rate and fire design;
2. Temporal and seasonal characteristics of the traffic;
3. Location of the fire and people inside the tunnel;
4. Availability of performances of the protection systems =.

2.1 Fire rate and fire design

Fire inside the tunnel is most likely caused by collision and vehicle defect triggering fire. The occurrence of fire in a tunnel is very rare but, in any case, it needs to be investigated as it can develop into a critical event characterized by the speed of fire development and the size of the fire which are influenced by the vehicle type, conditions and cargo materials, the airflow conditions in the tunnel, the performance of the protection systems and the structural safety features of the tunnel.

The fire frequency of observed occurrences related to the traffic and accidents in a given tunnel or road should be rated by number per vehicle multiplied by kilometres to account the effect of tunnel length and traffic density [6]. Thus, from the defined characteristics of the virtual tunnel and the data available from the literature, fire rate f_r is established. The average fire rate in Italy (occurrence rate of fire = 5.6 fires per 10^9 vehicles per km) that is available from the PIARC report "Experience with significant accidents in road tunnels" [7] is used to determine the fire rate of this specific tunnel.

$$f_r = \text{occurrence rate} \times \text{AADT} \times \text{length of tunnel} \times 365 = 1.54E-02 \text{ per year.} \quad (1)$$

If risk analysis is to be done in an existing tunnel, the records of the observed accidents on that specific tunnel and other related information that is essential for the analysis can be used. Observable data analysis can be carried out assuming the Poisson hypotheses and checking them with chi-square test when the Poisson distribution model is used for the said validation.

Heat release rate (HRR), the rate of heat generation and release by fire typically measured in watts, is a very important factor in the assessment of the severity of tunnel fires [8]. So, it is necessary to associate design fires of a given HRR for each scenario to properly simulate and describe the consequences of each and consequently account the number of fatalities and the complementary number of rescued out of the present people in the tunnel. While HRR varies for different conditions, the maximum HRR that can be observed on different types of vehicles can represent design fires. From the data gathered by PIARC [9], the maximum fire power observed ranges from 2.5 MW to 30 MW for different vehicles (passenger car, van, bus and lorry with burning goods) whereas the EUREKA HGV fire test indicated a peak power output of 100 MW to 120 MW for larger vehicles (HGV) with burning goods. The characteristic of the fire can be influenced by the type of material of the vehicle itself, the type of cargo, if it carries hazardous and flammable goods, and the amount of the hazardous goods. In Italy, ANAS [4] defined the different design fires related to the different types of vehicles which are obtained from the analysis of time series of the available accident datasets.

Since the HRR is different for different types of vehicles, the traffic composition in terms of vehicle type also has an influence on the fire rate. Hence, it is necessary to include the percentage shares of the traffic (light and heavy) in obtaining the probability of fire occurrence for different fire powers.

The proportion of the vehicles were assumed to be 85% for light and 15% for heavy vehicles (Fig. 1). This was taken on the basis of EU Directive 2004/54/EC [10] saying that heavy vehicles exceeding 15% of AADT requires to assess an additional risk by increasing the traffic volume of the tunnel in calculations. For this reason, it is assumed that there is an ideal situation where the traffic share of heavy vehicles does not exceed 15%. The values mentioned are in which the limits also used as a basic safety parameter in ANAS guidelines [4] which must be checked with real data for specific tunnels. The value for heavy vehicles that should be used in the analysis must be always greater than the recorded maximum traffic share of heavy vehicles for that particular tunnel.

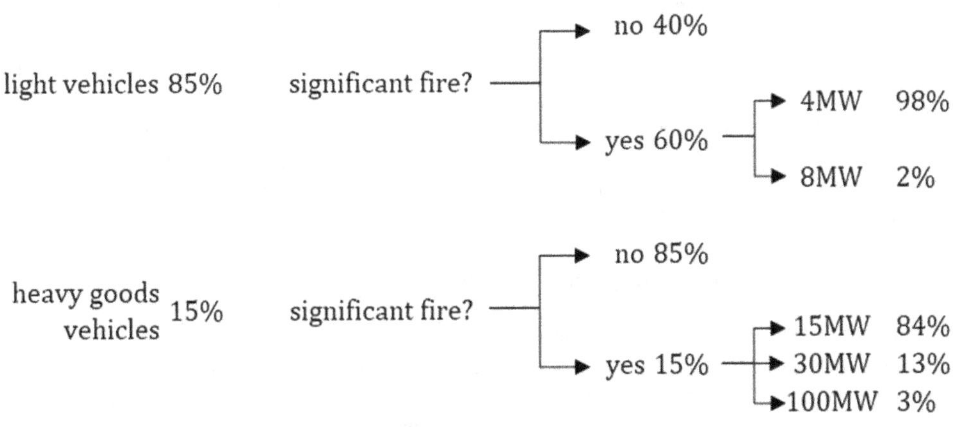

Figure 1: Distribution of heat release rate (HRR) according to ANAS guidelines.

The choice of heat release rate for fire scenarios is usually recommended by the working groups with representative members from the tunnel owners, fire brigades, regulators and consultants [9].

For this study report, the design fires are established based on the ANAS guidelines [4]. The probability of a fire accident of a given HRR, P(HRR), developed from the ANAS guidelines [4] are linearly interpolated and reclassified according to fire powers of 10 MW, 30 MW, 50 MW and 100 MW. Minor fires are the fires that do not create any injury or fatalities, or in general, the fires that are "insignificant" to risk analysis. Below you find the new probabilities of the standardized HRR fires defined (Table 1).

Table 1: Design fires and their probability of occurrence.

HRR (MW)	P(HRR)
Minor fire	7.18E-03
10	7.92E-03
30	2.52E-04
50	2.96E-06
100	7.41E-06

2.2 Temporal and seasonal characteristics of the traffic

PIARC [3] and Sandin et al. [6] both agreed that although most risk analysis used AADT, this number does not fit reality well since traffic, the number of vehicles, can vary seasonally and during the day, and therefore, this difference must be considered in the analysis.

A pattern of activities can be observed during the weekdays (Monday to Friday). Daily activities such as going to work or school are routinely made throughout the day. In Italy, working hours and school hours usually start from 8:00 or 9:00 and finish at 18:00 or 19:00.

Weekends (Saturday and Sunday) are typically spent for leisure activities like shopping and travelling with less commercial transportation. While these activities are more random compared to weekday activities, they are mostly performed within the typical working hours. It is also predicted that higher traffic will be observed in weekend than in weekday due to the reduced public transport availability.

Therefore, the observations on people's activities and the statistics from different reports became the basis for considering different days of the week and different time periods of the day in identifying the scenarios. The traffic flow during the daytime (from 6:00 to 20:00) is assumed to be higher compared to the traffic flow during the night (from 20:00 to 6:00). Consequently, the probability of an accident occurring in different periods is expected to be proportional to their traffic flow variation.

2.3 Location of the fire and people inside the tunnel

According to the PIARC report "Experience with significant accidents in road tunnels" [7], the collision rates were significantly higher in the portal area (10 m before/after tunnel portal) and in the entrance area (10–150 m inside the tunnel) compared to the interior zone (more than 150 m from both portals). When entering a tunnel, drivers experience a sudden change in illumination causing eye confusion which could trigger their driving behaviour that might result to an accident.

Although many studies reported that most of the ordinary accidents (collisions) in tunnels occurred in the entrance, only the interior zone and the exit area of the tunnel were included in the scenarios for this report. Rather than taking all the accidents (collisions) into account, the interest of this report is focused on accidents involving fire. And assuming that the vehicles downstream the accident will not be affected since they are able to exit the tunnel while the vehicles upstream might be directly exposed to the hazard since the area of accident will block the incoming vehicles, therefore for the accidents, involving fire, that will occur in the entrance area, very few vehicles will be directly exposed to the hazard and in most cases the chance of having an injury or a fatality is very low and almost zero. And despite having a high probability of accident occurring in the entrance area, since we analyse the scenarios using their quantum of risk (scenario contributions to the expected value) which is the product of the probability of the scenario and the quantified consequence in terms of fatalities, the contribution of the scenarios in the entrance area would be very low approaching to zero. Having said that, the two worst locations that seem representative for different possible locations of the fire inside the tunnel are identified as the interior zone and the exit area.

To determine the fatalities resulting from each scenario, the number of exposed units must be defined. The exposed units are not equivalent to the number of people present inside the tunnel as this will underestimate the individual risk caused by the hazard. Hence, since the vehicles downstream the accident can continue to exit the tunnel, only the vehicles queued upstream the accident are counted.

A simple queue model described by Persson [11] was used to determine the exposed units that needs to be evacuated during a fire incident. He used a time period of 2 minutes before the warning system is activated following the fire accident, and 1 minute before all vehicles notice the warning signs and stop. The vehicles used in the calculation have an average length of 3.5 m and are at least 1.5 m apart with four people inside each vehicle. The average daily traffic used is 15,000 vehicles per day, and the vehicles run at an average speed of 70 km/h.

Among the major tunnel accidents in Europe, the Monte Bianco (1999), Tauern (1999) and Gotthard (2001) tunnels, the number of persons in the tunnel reported ranges from 60 to 180 people with fatalities ranging from 11 to 39 people. Thus, for this study report, the number of exposed units is identified and limited as the minimum between the calculated "total number of people upstream the accident" and 50 (for unidirectional tunnel) or 100 (for bidirectional tunnel) (Table 2).

Table 2: Exposed units in an event of fire inside the tunnel.

Time of the day	Traffic flow (veh/h)	distance from the entrance (m)	unidirectional tunnel		bidirectional tunnel	
			total number of people inside the tunnel (upstream the accident)	**exposed units**	total number of people inside the tunnel (upstream the accident)	**exposed units**
daytime	857.14	501	198	**50**	198	**100**
		334	190	**50**	224	**100**
nighttime	300.00	501	74	**50**	74	**74**
		334	74	**50**	142	**100**

The derivation of the probabilities of the mutually exclusive initiating events are described in Fig. 2. The probability of a fire occurring with a given power in terms of heat release rate are coupled with a series of assumptions on the major key factors mentioned earlier. The factors included here are the ones that can influence the hazard scenario, independent from the impact of the safety characteristics of the tunnel such as the performance of the illumination and ventilation systems in case of fire.

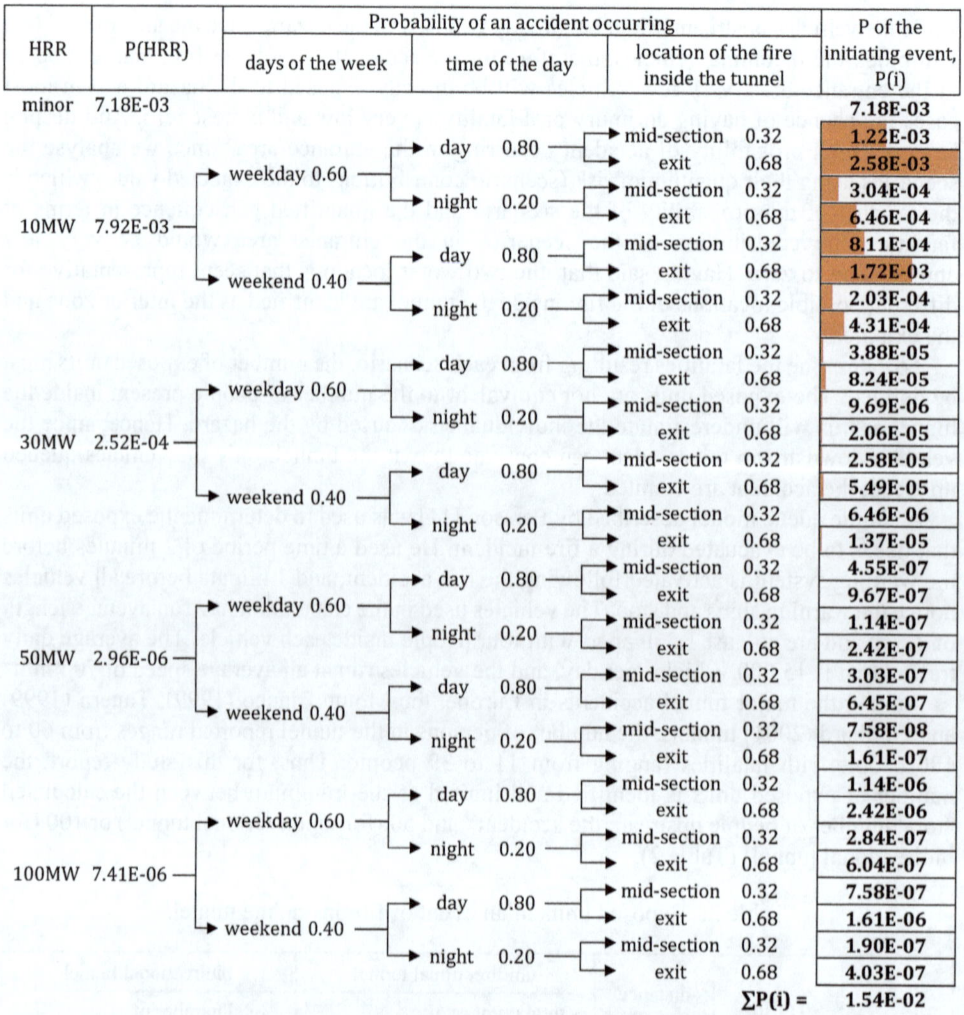

Figure 2: Probabilities of the initiating events.

2.4 Availability of performances of the protection systems

Opacity or reduced visibility and the inhalation of toxic substances due to smoke emission influences the walking speed of the exposed units and therefore the evacuation time. According to PIARC [9], there is a linear correlation between the smoke dependent visibility measured by the optical density and the concentration of CO_2. For a given visibility of 10 m, the optical density must not exceed 0.13 m^{-1} which means that the CO_2 concentration must be around 0.05% to 0.3%. But the maximum values of 2% to 16% CO_2 were observed on different burning materials from the vehicles (e.g., seats, tyres, fuel, etc.) therefore intense smoke production might result to a very reduced visibility downstream from the fire load [9]. In addition, smoke gases in the tunnel do not only reduce the visibility but also creates health danger from possible toxicity especially from the concentration of carbon monoxide [9]. A sufficient amount of carbon monoxide inhaled by a person can cause headache, weakness, confusion, shortness of breath, and eventually can cause loss of consciousness that can lead to death with continuous exposure [11]. According to studies, the principal causes of death from a fire are due to carbon monoxide poisoning followed by carbon dioxide poisoning and oxygen deficiency [12].

Emission of toxic gases are also observed on the pavement flooring (bituminous conglomerate/asphalt) of the tunnel creating an additional impact on the severity scenario. The asphalt begins to emit toxic substances after 5 minutes of exposure to temperatures of 428°C to 530°C and catches fire after 8 minutes. The combustion of the asphalt during tunnel fire worsens the temperature and the visibility making it more difficult for the exposed units to escape. As for concrete pavement, it has an advantage in an event of fire as it does not generate heat and do not emit additional smoke, and have a better ability to maintain its characteristic in case of fire [13].

The concept of fractional effective dose (FED) can be used to determine the time until one person is incapacitated and die in the tunnel because of toxic inhalation using the two main parameters that should be considered when analysing the effects of toxic to a body – the concentration of the toxic substance and the duration of the exposure [11]. Purser and McAllister [14] said that when FED = 1, half of the population is expected to be incapacitated while and FED = 0.3 can cause incapacitation for the 11% of the population.

Based on the provisions of EU Directive 2004/54/EC [10], a tunnel with 501 m length and AADT > 2,000 vehicles per lane per day is required to have illumination and the ventilation systems. These protection systems are the two major complex protection systems that will help the exposed units to face the opacity and toxicity due to the forming smoke. And the parameters that define the performance of the protection systems concerns the technical design of the supplier and the sense of responsibility of the tunnel authorities (maintenance and recovery), and not of the risk analyser. Each initiating event is coupled with the availability of the performance of the two fundamental protection systems, giving the probability of the scenario. The "availability" described is the availability and the unavailability of the protection systems in case of an accident resulting to fire which can affect the ability of the people inside the tunnel to self-rescue.

Illumination performance and availability can be measured in terms of the total time in a year that the lighting system is able to provide illumination of a given level of lighting (lux, lumen, candela) during normal situation and a minimum level of visibility for users to allow them to leave the tunnel in their vehicles in case of electrical failure or to provide light to people walking to reach an escape route during an emergency.

As stated in the ANAS guidelines [4], emergency lighting during an electrical failure should guarantee an illumination of 1 cd/sqm for the entire tunnel for at least 30 minutes. While the safety lighting used for the evacuation of the users, in case of emergency, must illuminate the walkway with an average illuminance of 5 lux and a minimum illuminance not less than 2 lux, and with a minimum range of 90 cm.

Sandin et al. [6] performed a series of test to calculate the time it takes for a person to reach the closest unobstructed exit (100 m) and it shows that for a given unimpeded walking speed of a person with reducing visibility, the time required to evacuate increases. The unimpeded walking speed of 1m/s with visibility of 10 m, 5 m, 2 m, and 0.5 m resulted to 100 s, 100 s, 152 s, and 500 s, respectively. Moreover, the evacuation time also increases with increasing people density.

Ventilation performance and availability can be measured in terms of the total time in a year that the ventilation system is able to provide adequate air quality during normal situation and maintain a minimum level of toxicity during emergency allowing people to self-rescue.

In case of fire, the ventilation system must disperse the thermal energy generated by the fire, manage and control the smoke propagation, and exhaust the toxic and flammable substances.

Persson [11] investigates the time before the exposed people at different positions from the fire reach the critical values of FED. The scenario resulting from HGV with HRR = 20 MW shows that the people located at 10 m, 20 m, 30 m, and 40 m from the fire will not experience incapacitation before they complete the evacuation. The same findings happen for the scenario resulting from HGV with HRR = 120 MW although the critical values of FED are reached after 40 minutes. In case of pool fires, worst consequences are predicted, and shorter times are calculated to reach FED = 1.

Sandin et al. [6] also makes the same analysis using FED to determine the time to incapacitation due to the combination of carbon monoxide (CO), carbon dioxide (CO_2) and dioxide (O_2), and incapacitation due to heat. A combination of 3% CO_2, 4,000 ppm CO and 17% O_2 will reach FED = 1 at 220 s while a more disadvantageous combination with 8% CO_2, 10,000 ppm and 9% O_2 will reach FED = 1 at 35 s.

3 SCENARIO DESCRIPTION AND SIMULATION

3.1 Aims of simulations

Basically, the scenarios simulations aim to numerically calculate the number of estimated fatalities coupled with any design scenario. Each one of the above scenarios is also coupled with the estimation of its probability of occurrence estimated with the parallel established procedure based on the tree diagram.

Each design scenario does correspond to one specific branch of the tree diagram. Each simulation provides our estimated value of the number of fatalities that can be compared with the number of people present in the considered tunnel and out of them with the number of the ones that are inside the distance of influence (potential damages) from the fire accident location.

The last number mentioned above is the estimation of the so-called exposed units number. The ratio between the number of fatalities and the exposed units is a first rough estimate of the mortality rate of the given scenario.

The quantum of risk estimate need the corresponding value of the estimated probability of the scenario itself. The simulations include the protective effect of the mandatory protection systems active in a tunnel according to length, volume and composition of the traffic.

The simulator does create exactly the geometrical and exposure conditions of the specific tunnel in which will take place the initiating event of fire accident and consequent processes of toxicity diffusion and exposed units evacuations (Fig. 3).

3.2 Necessary INPUTS and expected OUTPUTS

Location coordinates and initial walking speeds of the exposure units participating to a given scenario have to be available. The geometrical, architectural and structural condition of the Tunnel have also to be available.

The heat release rate curve and the performance levels of the available protection systems have to be specified.

The number of fatalities for each given scenario are then expected together with the details of the available safe egress time (ASET) and the required safe egress time (RSET) for each one exposure unit.

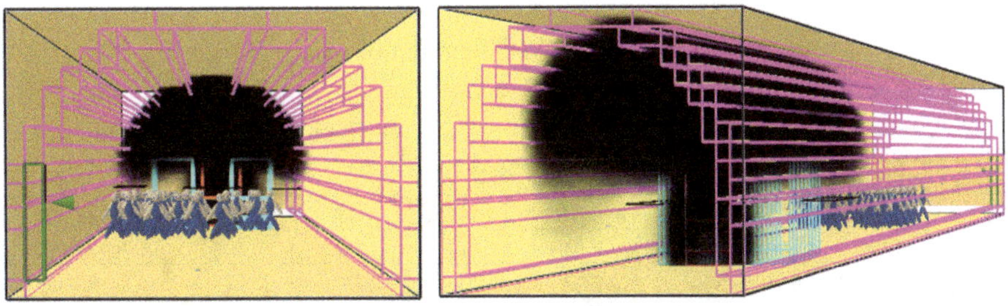

Figure 3: Two views of the exposed units locations, each of them walking with an assumed walking speed at the same moment in time. All the data are recorded and available for processing purposes from the fire initiating event time to the final picture accounting for number of fatalities and rescued people (exposed units). The above data are relevant for the aim of optimizing the risk-based design model.

4 SCENARIO PROBABILIZATION AND RISK QUANTA GU@LARP MODEL

In total, 128 representative scenarios are identified from the combination of the different conditions and factors that are most relevant to the risk analysis of this tunnel. Now, the methodology used for scenario probabilization is described on the following paragraphs.

The scenario tree diagrams of the initiating events are the computational tools allowing us to measure the quantum of risk of each simple scenario by considering the impact of the performance of the protective systems, the illumination and ventilation systems, on the hazard flow in terms of the lack of visibility and the lack of breathability (Fig. 4).

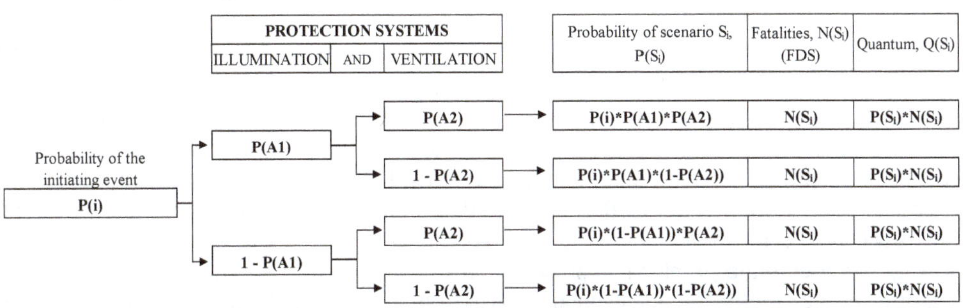

Figure 4: Scenario tree diagram of an initiating event.

Each tree produces a complete group of mutually exclusive scenarios originating from the same initiating event and therefore share the probability of the initiating event among them. The probability of each scenario $P(S_i)$ being given by the corresponding simple tree branch is the product of the probability of the initiating event P(i) and the complex of the performance of the illumination and ventilation systems. The quantum $Q(S_i)$ is then the result of the probability of scenario multiplied by the corresponding number of fatalities $N(S_i)$ given by the fire dynamic simulator (FDS).

Finally, for the risk-based design, the quantum of risk of each scenario can be compared and ranked according to its severity – the higher the quantum, the higher the risk. Table 3 shows the calculations for the unidirectional virtual tunnel.

Table 3: Ranked quantum of risk.

	Scenario, S_i*		Probability of the Scenario, $P(S_i)$	Fatalities, N_i	Quantum, Q_i
1	5	10, WD, D, E, GI, GV	1.74E-03	5	8.72E-03
2	6	10, WD, D, E, GI, PV	5.81E-04	15	8.72E-03
3	1	10, WD, D, M, GI, GV	8.21E-04	6	4.92E-03
4	21	10, WE, D, E, GI, GV	1.16E-03	3	3.49E-03
5	21	10, WE, D, E, GI, GV	1.16E-03	3	3.49E-03
6	2	10, WD, D, M, GI, PV	2.74E-04	11	3.01E-03
7	7	10, WD, D, E, PI, GV	1.94E-04	12	2.33E-03
8	17	10, WE, D, M, GI, GV	5.47E-04	4	2.19E-03
9	13	10, WD, N, E, GI, GV	4.36E-04	4	1.74E-03
10	14	10, WD, N, E, GI, PV	1.45E-04	12	1.74E-03
11	8	10, WD, D, E, PI, PV	6.46E-05	20	1.29E-03
12	18	10, WE, D, M, GI, PV	1.82E-04	7	1.28E-03
13	9	10, WD, N, M, GI, GV	2.05E-04	5	1.03E-03
14	23	10, WE, D, E, PI, GV	1.29E-04	7	9.04E-04
15	3	10, WD, D, M, PI, GV	9.12E-05	8	7.30E-04

Among the 128 scenarios identified, the 15 scenarios with highest quantum of risk are shown on table comprising the 86% of the total quantum of risk of the tunnel analyzed. It is evident that the *scenario 5* with HRR = 10 MW occurring during the weekday (daytime) at the exit zone of the tunnel with good illumination and good ventilation gives the highest contribution to the total risk followed. Despite having a good performance of both the illumination and ventilation system that, the scenario 5 gives the quantum of risk due to its high probability of occurrence.

In the risk-based design, these information regarding the possible scenarios in case of fire is important to determine the compliance with the risk acceptability/tolerability criteria, following the ALARP principle or "As Low as Reasonably Practicable", to check if the minimum level of safety is guaranteed in the tunnel. When the risk is intolerable, risk reduction must be made regardless of the cost; when the risk is tolerable, it is necessary to check if risk reduction is not further possible due to unbalanced cost vs benefits; when the risk is acceptable, no additional measure is required.

Fig. 5 presents the graphical representation of the societal risk (frequency–number of fatalities curve) of the tunnel evaluated based on the criteria established by the Italian Legislative Decree 264 [15]. The acceptable risk corresponding to the abscissa fatality N = 1 is 10^{-4} per year and the abscissa fatality N = 100 is 10^{-6} per year. Whereas the tolerable risk corresponding to N = 1 is 10^{-1} per year and N = 100 is 10^{-3} per year.

The "tolerability risk line" and "acceptability risk line" correspond to the exceedance probability distribution function according to the Gu@larp model. The Gu@larp density function is $g(N) = Gu/N^2$ and the corresponding exceedance probability is $G(N) = Gu/N$ where Gu is the risk indicator which is the value of the ordinate at N = 1.

The "risk line" shows the exceedance probability curve of the societal risk in the tunnel. For a given number of fatalities N in the abscissa, the probability of exceedance is equal to the result of back-cumulation procedure of all the probabilities of the scenarios with higher than or equal to that threshold N. The area below each staircase of the exceedance probability curve is then the summation of the quantum E of all scenarios having the same number of fatalities N.

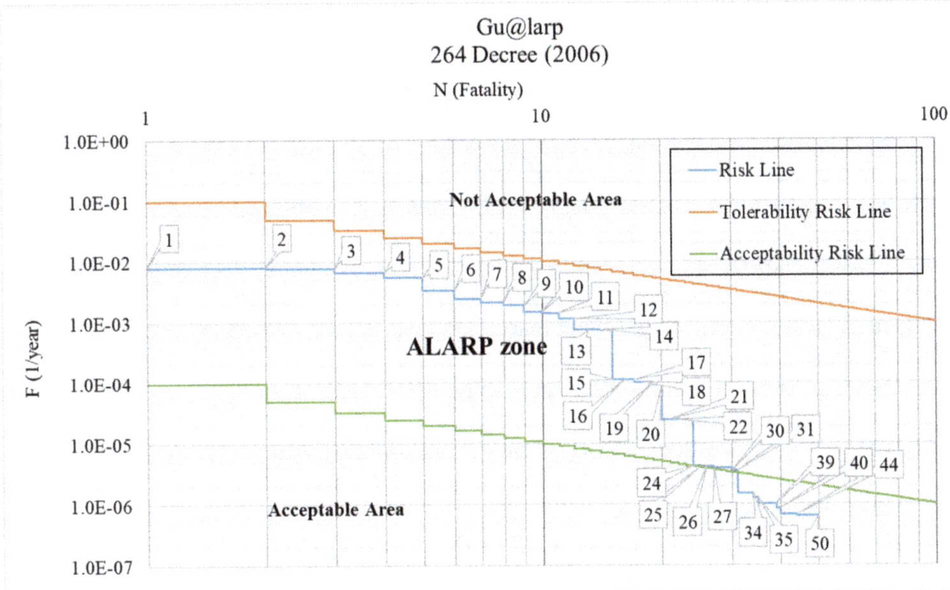

Figure 5: ALARP criteria based on the Italian Decree 264.

The question of risk-based design is whether to improve the scenario with the highest risk or the scenario with the highest fatalities. In reality, people would be more concerned about the scenarios with the highest fatalities more than the scenarios with low fatalities. More often, the probability of scenarios with high fatalities are extremely low which also results to low quantum of risk. And since higher quanta indicates greater space for improvement and therefore, improving "high-risk" scenarios might largely contribute to the reduction of societal risk rather than improving "high-fatalities" scenarios.

5 CONCLUSIONS

The Gu@larp risk quantum model is fully respectful of the concept of risk quantitative measure for forensic and societal purposes as has been for the first time but very clearly stated in the 1969 UK court judgment by the judge Lord Asquith.

The individual risk indicator is firstly clearly stated in relationship with the concept of quantum of risk of a given scenario out of a well identified specific corresponding design scenarios.

From the quantum of risk coupled with the all necessary design scenario the requested risk indicators for the compliance with the acceptability and tolerability risk criteria are immediately calculated.

REFERENCES

[1] Guarascio, M., As low as reasonably practicable. *Transactions on The Built Environment,* vol. 206, WIT Press: Southampton and Boston, 2021.

[2] Guarascio, M., Italian risk analysis for road tunnels. PIARC Technical Committee C3.3 Road tunnel operation, pp. 209–223, 2008.

[3] PIARC, Road analysis for road tunnels. PIARC Technical Committee C3.3 Road tunnel operation, 2008.

[4] SpA, ANAS Linee Guida per la progettazione della sicurezza nelle Gallerie Stradali secondo la normativa vigente, 2009.
[5] Guarascio, M., Lombardi, M., Rossi, G. & Sciarra, G., Risk analysis and reliability based design in tunnel fire safety. *Transactions on The Built Environment*, WIT Press: Southampton and Boston, pp. 575–584, 2009.
[6] Sandin, K., Grenberg, K., Husted, B.P., Scozzari, R., Fronterrè, M. & Ronchi, E., *Verification and Validation of the ARTU (Tunnel Fire Risk analysis) Tool*, Department of Fire Safety Engineering, Lund University, 2019.
[7] PIARC, Experience with significant incidents in road tunnels. Technical Committee 3.3 Road Tunnel Operations, 2016.
[8] Ingason, H., Li, Y.Z. & Lönnermark, A., *Tunnel Fire Dynamics*, Springer Science +Business Media: New York, 2015.
[9] PIARC, Fire and smoke control in road tunnels. PIARC Committee on Road Tunnels Operation (C5), report 05.05.B, 1999.
[10] EU, EU Directive 2004/54/EC of the European Parliament and of the Council of 29 April 2004 on minimum safety requirements for tunnels in the Trans-European Road Network, 2004.
[11] Persson, M., *Quantitative Risk Analysis Procedure for the Fire Evacuation of a Road Tunnel: An Illustrative Example*, 2002.
[12] Gormsen, H., Jeppesen, N. & Lund, A., The causes of death in fire victims. *Forensic Science International,* **24**(2), pp. 107–111, 1984.
[13] Domenichini, L., La Torre, F. & Caputo, F.J., Impiego delle pavimentazioni in calcestruzzo nelle gallerie stradali. *Strade and Autostrade,* **10**(60), 2006
[14] Purser, D.A. & McAllister, J.L., Assessment of hazards to occupants from smoke, toxic gases, and heat. *SFPE Handbook of Fire Protection Engineering,* 5th ed., ed. M.J. Hurley, Springer Science+Business Media, pp. 2308–2428, 2016.
[15] D. Lgs 05 ottobre 2006, n. 264: Attuazione della direttiva 2004/54/CE in materia di sicurezza per le gallerie della rete stradale transeuropea. (2006). *Sicurezza gallerie stradali*, 2006

MANAGING HOMELESS PATIENT RISK IN A U.S. HEALTHCARE SYSTEM

ERIC S. KYPER[1], MICHAEL J. DOUGLAS[2] & LUKAS S. LEES[2]
[1]College of Business, University of Lynchburg, USA
[2]Lombardo College of Business, Millersville University, USA

ABSTRACT

Homeless patients are typically treated in emergency departments (EDs). EDs have the capability of treating a large variety of issues. Financial, staffing, and physical resources are often strained. ED visits for access to free food and hygiene facilities are common. Other times the ED is used to escape bad weather conditions. These visits strain the waiting room, use valuable staff resources to resolve non-critical care issues, and ultimately cost the hospital out-of-pocket money because the hospital receives no revenue. To improve ED performance, we studied a patient re-routing system for the homeless that better places these patients into hospital owned services that are separate from the ED. For example, transportation to a free health clinic (not walk-in) with an assigned primary care physician improves patient health for issues like diabetes. The hospital also owns a pharmacy clinic that now works closely with the free health clinic to provide medications and instructions for homeless patients. The program has been successful in reducing ED costs and hospital admissions for acute homeless patients.

Keywords: homeless, emergency department, risk management, overcrowding, re-routing, free clinic.

1 INTRODUCTION

Homelessness in the United States is a multi-faceted epidemic. Simultaneously homelessness is an economic problem, a housing problem, and a transportation problem. Widespread unemployment, poor living conditions, and walking as a primary means of transportation create societal issues involving every citizen in the U.S. This paper focuses on the healthcare issues involving homeless people and emergency department services in regional healthcare systems in the United States. According to the National Hospital Ambulatory Medical Care Survey [1], a homeless person is almost five times more likely to visit the emergency department (ED) than the general population.

A homeless person will use total healthcare less than an average individual [2] but instead uses the ED as their primary care facility. Difficult living conditions for the homeless lead to healthcare requirements that typically fall on ED services. The ED is an attractive option for the homeless because no appointment is necessary. However, the data analysed demonstrates that many ED visits by the homeless are for low acuity issues. These are often related to diet, hygiene, and lack of access to routine care of treatable conditions like diabetes. By law, a hospital must assess and treat all patients, regardless of the patients' ability to pay. Hospitals absorb almost all costs related to treating homeless patients. One option for the hospital to manage homeless care and costs is to enter patients into a risk management system designed to reduce costs and increase homeless patients' overall health.

In this paper, we analyse a hospital system that identifies homeless patients and re-routes them to more effective and lower cost treatment options within the hospital network. Homeless families with children and homeless with disabilities are particularly vulnerable [3] and ideally should have prioritized care. The re-routed patients are placed in the Excalibur system. The Excalibur system provides patients and alternative to using the ED as a primary care facility by giving the patients access to a free clinic with primary care physicians and a pharmacy, both owned by the hospital network.

WIT Transactions on The Built Environment, Vol 214, © 2022 WIT Press
www.witpress.com, ISSN 1743-3509 (on-line)
doi:10.2495/SSR220041

2 RESEARCH METHODOLOGY

The research methodology used in this paper was to analyse secondary data. The data was collected through normal hospital operations and then manually extracted from hospital records. Identifying patient information was removed to comply with federal guidelines (HIPPA). A year worth of microdata from a regional ED from 2013 to 2014 was studied. The data consists of 6,288 homeless patient visits.

2.1 Data

The data included:

- Patient ID – an assigned number for each patient to remove identifying information.
- Patient status – a binary number to represent if the patient is in the Excalibur or not in the system at the time of visit to the ED.
- ESI acuity code [4] – a number between 1 and 5, where 1 is the highest priority and case severity, and 5 is considered non-urgent. Some patients do not quality for an acuity code. Those cases were designated as ACUNASSIGNED. Examples of triage nurses using ACUNASSIGNED include homeless coming in the ED to ask what time it is or if they have any free food. All ACUNASSIGNED patients were discharged. In addition, false reports of abuse were also assigned ACUNASSIGNED.
- Status – a binary number represented if the patient was admitted to the hospital or discharged from the ED.

2.1.1 Research questions

Research question 1: Are there visitation differences in ESI Acuity Level between patients in Excalibur versus out of Excalibur? Research question 2: Does the re-routing of patients using Excalibur result in financial savings for the hospital?

2.1.2 Analysis

The primary analysis for research question 1 was to visualize the data, conduct a crosstab to describe the data and conduct a chi-square test using the patient status and ESI acuity code. Fig. 1 and Table 1 demonstrates that Excalibur is successful in reducing ED visits by the homeless across all acuity levels.

A chi-square test was conducted to see if the proportions of patients based on the ESI acuity levels were different between Excalibur and non-Excalibur patients (Table 2).

The chi-square test confirm statistically significant differences between the patients in and out of the Excalibur system. The Pearson chi-square test reports a p-value of .029 meaning at the .05 level, there is significant evidence of association between the two variables (ESI Acuity and pre/post Excalibur). However, when controlled for patient admission or discharge the chi-square test is statistically not significant for admitted patients. This supports the idea that if a patient is sick enough to be admitted to the hospital, then the free clinic and pharmacy is not going to make a difference. The ED discharge rates demonstrate reducing the number of ACUNASSIGNED patients, by giving them alternatives to going to the ED, makes a significant difference.

2.1.3 Financial analysis

Our second research question is related to financial savings. Fig. 2 demonstrates the substantial savings that result from the Excalibur program. Financial savings are necessary for Excalibur to be sustainable.

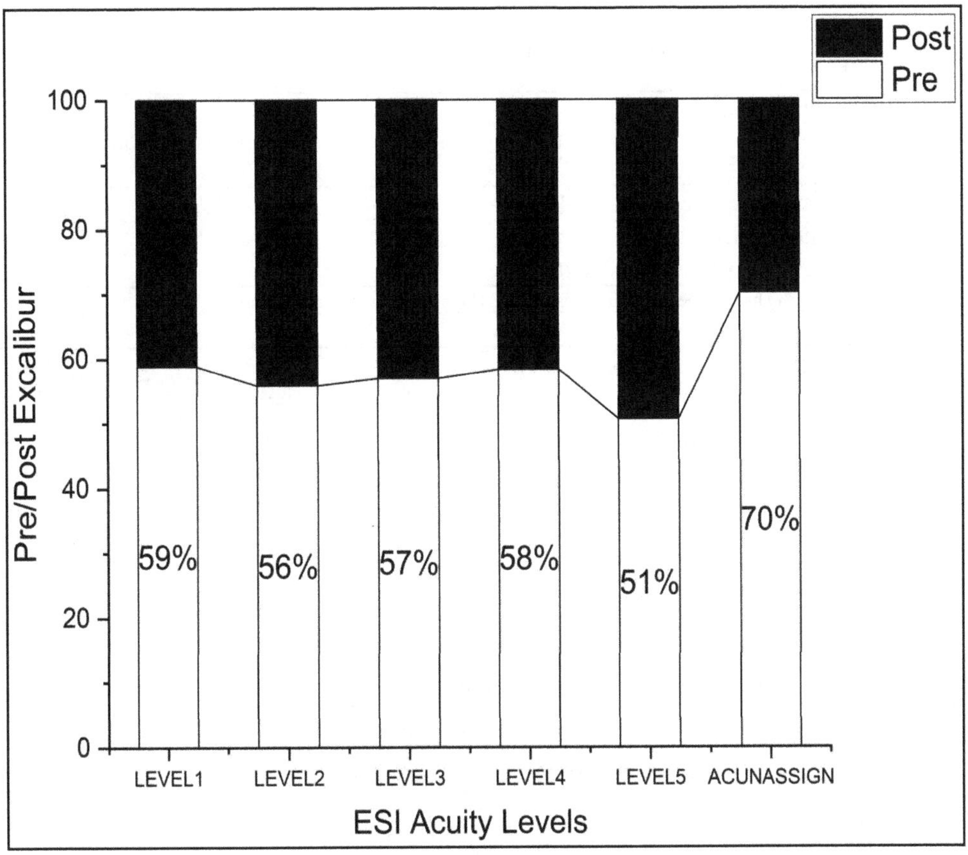

Figure 1: Percentage of visits by acuity level pre and post inclusion in Excalibur.

Table 1: Cross tabulation of visit counts by acuity level and inclusion in Excalibur.

		Pre/post Excalibur		Total
		Non-Excalibur patient	Excalibur patient	
Acuity code	LEVEL1	30	21	51
	LEVEL2	1,019	805	1,824
	LEVEL3	1,595	1,206	2,801
	LEVEL4	681	487	1,168
	LEVEL5	191	186	377
	ACUNASSIGN	47	20	67
Total		3,563	2,725	6,288

Table 2: Results for chi-square tests.

Chi-square tests	Value	df	Asymptotic significance (2-sided)
Pearson chi-square	12.430[a]	5	.029
Likelihood ratio	12.583	5	.028
Number of valid cases	6288		

[a]0 cells (.0%) have expected count less than 5. The minimum expected count is 22.10.

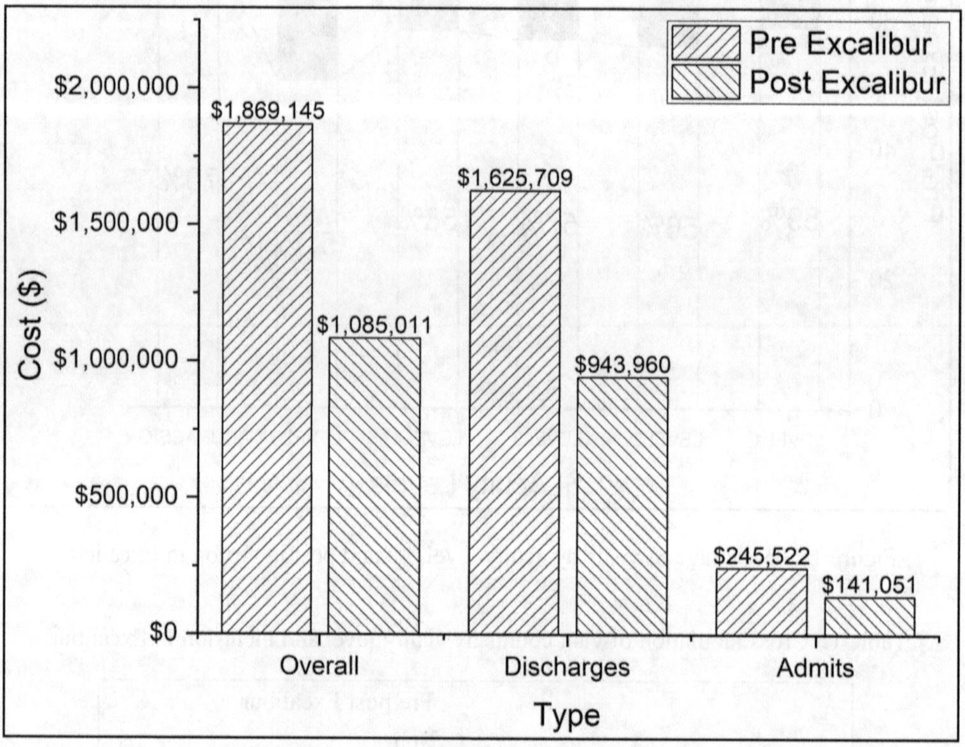

Figure 2: First year savings due to the Excalibur system.

The total savings for the first-year amount to $784,134 overall with discharge savings of $681,749 and admit savings of $104,471.

3 IMPLICATIONS

Several benefits are expected from a system like this. From the hospital's perspective, the cost savings are upfront and substantial. This system costs around $100,000 to implement the first year. With first year savings of $784,134 the return in investment is excellent. Continuing costs to maintain Excalibur are estimated at $50,000 per year. If effective, we expect the cost savings per year to decrease, with the goal of getting every homeless person in the Excalibur system.

From the public perspective ED wait times will improve, all else being equal. Valuable resources are freed up to attend to high acuity cases benefiting everyone. Homeless people are better served in this system with access to an appointed primary care physician who can help manage their health. Homeless shelters and the like benefit from a healthier population and the knowledge they can direct individuals to proper care facilities. Society also benefits. Lower hospital costs help fight healthcare inflation. A healthier homeless community is less likely to contract or spread diseases such as tuberculosis and hepatitis.

4 FUTURE EXTENSIONS

Integrating a transportation component to coordinate patient pick-ups/drop-offs between shelters such as the Salvation Army and treatment facilities is the next step. Lack of transportation is the biggest challenge in getting homeless people the care they need.

Risk management through simulation is also possible. Hospital systems need to spend money to manage the monetary risk of non-paying homeless patients. The amount of money to spend can be analysed with simulation. While we do not want to change funding levels to see the differences in homeless level patients to the ED in real life, a simulation will allow for optimizing funding levels for the Excalibur system. The goal is to have close to 0 non-Excalibur patients. But the irony for management is that if every patient was an Excalibur patient, then there will be no cost savings of the program in the way the hospital system accounts for the homeless/non-payers. Even though the hospital system would be taking more than a one-million-dollar charge for not having the system at all. The benefits though will outweigh the low-cost savings.

REFERENCES

[1] U.S. Department of Health and Human Services, QuickStats: Rate of emergency department (ED) visits, by homeless status and geographic region. National Hospital Ambulatory Medical Care Survey: United States, 2015–2018, 2020.

[2] Zhou, R.A., Baicker, K., Taubman, S. & Finkelstein, A.N., The uninsured do not use the emergency department more-they use other care less. Health Affairs, **36**(12), pp. 2115–2122, 2017.

[3] Shinn, M. & Richard, M.K., Allocating homeless services after the withdrawal of the vulnerability index service prioritization decision assistance tool. *American Journal of Public Health*, **112**(3), pp. 378–382, 2022.

[4] Hong, R., Sexton, R., Sweet, B., Carroll, G., Tambussi, C. & Baumann, B.M., Comparison of START triage categories to emergency department triage levels to determine need for urgent care and to predict hospitalization. *American Journal of Disaster Medicine*, **10**(1), p. 9, 2015.

SECTION 2
RISK IN CONSTRUCTION AND MANAGEMENT

PEDESTRIAN MOBILITY IN THE PROXIMITY OF CONSTRUCTION SITES: AN APPROACH TO ANALYSE AND IMPROVE THE PEDESTRIAN EXPERIENCE

ALESSIO MORICCA & VEDRANA IKALOVIC
Graduate School of Engineering Junia – HEI, Lille Catholic University, France

ABSTRACT

Pedestrian mobility is the only mobility that is included in all types of itineraries, while pedestrians are one of the most vulnerable users of the road. In the Hauts de France region, which has been transforming from an industrial into an IT and service-oriented region, it is possible to denote a massive presence of construction sites that cause numerous issues to the local community in their proximity. Literature analysis shows that researchers often underestimate the importance of pedestrians, hence, one of the problems which is analysed in this study is the mobility in construction site proximity. The purpose was to analyse the comfort of the sidewalks and the pedestrian behaviour. The analysis of pedestrian mobility is conducted in the regional capital Lille including (1) an analysis of the history of Lille's Catholic University and the territory using GIS; (2) a CAWI survey to study the attitude of pedestrians and their opinions about construction sites; and (3) an observation of three sidewalks located in proximity of construction sites. Observations and questionnaires show that residents are willing to accept construction sites and adapt to the changes, showing particular interest in the construction projects and their progress. In their interest lies an opportunity for a better integration of construction sites within the neighbourhood. The study concludes that better communication between stakeholders using real-time plans of temporary dynamics, would bring specific improvements to different types of sidewalks and streets.

Keywords: construction sites, pedestrian mobility, road safety.

1 INTRODUCTION

New constructions and renovation processes of existing buildings, in general are bearers regional development, transformation, and economic growth [1]. However, the negative effect of construction is inevitable and includes noise, environmental pollution, dust, traffic issues, etc. Noise issues and their impact on the neighbourhood are widely described in many research papers: Ng has been conducting experiments on residents of student dorms [2], while Hong et al. created a rating system to define the impact of noise pollution on residents [3]. The problems of the environment and pollution are commonly analysed using traditional measurement methods and cost analysis [4], [5] and they are regulated by local and European laws [6]. During the initial observations in Lille, important issues regarding traffic had been noticed, especially for pedestrians and cyclists, and particularly for persons with reduced mobility (PRM) who must pass by the construction site. An attempt to analyse this issue has been made in an occupational safety journal where the problem has been analysed in relation to construction site employees [7]. Also, pedestrian traffic is often analysed by scientific papers and technical booklets to determine pedestrian comfort [8]. Finally, issues related to PRM mobilities and the effects of temporary architecture on pedestrian mobility were subjects of separate studies [9], [10]. Therefore, literature review has shown only a very limited analysis of the impact of construction sites on a pedestrian flow. This study bridges that gap discussing issues of pedestrian mobility within the sustainable transportation framework, and with the 11.2 goal of the United Nations sustainable development in mind [11].

WIT Transactions on The Built Environment, Vol 214, © 2022 WIT Press
www.witpress.com, ISSN 1743-3509 (on-line)
doi:10.2495/SSR220051

Pedestrian mobility composes our everyday lives. Almost every trip includes at least one part of mobility which needs to be done on foot. Examples of this phenomenon can be driving, the necessity to walk to and from the parking lot or taking public transport, the need of reaching the public transport stop, etc. The current approach of the European Union (EU) classifies pedestrians as vulnerable road users (VRU) and identifies them as one of the seven main focuses of the EU Road Safety Policy Orientations 2011–2020 [12]. Pedestrian mobility has also been strongly encouraged in the last years as a mitigation to the current climate change crisis [13].

In parallel, newly constructed buildings as well as renovated historical buildings are developed to provide comfortable spaces with high efficiency that meet sustainability goals. However, a large-scale construction sites and renovation projects that are common in historical cities (such as Lille), in multiple cases cause a long-term impact on daily life of local communities through the sound, vibrations, dust, and traffic pollution. The construction companies often overlook the impact on the community, especially on pedestrians in the proximity of their construction sites. Therefore, this research is an effort to address that issue and to face modern challenges of pedestrian mobility [14].

2 THE CITY OF LILLE AND THE CAMPUS OF THE CATHOLIC UNIVERSITY

This study explores Lille, a city located in the north of France, in the Hauts-de-France region, counting 236,400 inhabitants and an area of 446.7 km^2 divided into 12 quarters. The city is inscribed in the Metropole Europeenne de Lille (MEL) accounting 1,174,273 inhabitants and 672 km^2 [15]. It is a historical city, with industrial past, where large scale construction sites regenerate existing city tissue. The city has developed together with The Catholic University of Lille, located in its heart, which brings 15,000 of students [16] to the Vauban-Esquermes neighbourhood and its 19,335 inhabitants [17].

2.1 The city of Lille

The history of the city begins in the year 620 and the word “L’Isle” appears for the first time in a document dated 1066. The town begins to develop next to the port, located in the current “Vieux-Lille” neighbourhood. Lille develops strongly throughout the centuries, especially thanks to its strategic position between the Netherlands, Flanders, and the flow of business connections [18].

The development of the Vauban-Esquermes area can be dated approximately to the 17th century when Lille becomes a French city. Soon after the overtaking of the Lille by the French, king Louis XIV ordered the renovation of the fortifications and the construction of a Citadelle, which remains the city core until today. The name of the neighbourhood derives from the engineer Vauban, who designed fortifications in the Citadelle [19].

The position of the city of Lille since its foundation makes it an important transportation hub in central Europe and provides an important base for urban development: connectivity remains one of the city’s main characteristics. The metropolitan area of Lille is the fourth biggest agglomeration in France causing it to constantly develop and extend its city borders. New roads and buildings are currently being designed to be friendly for pedestrian, cyclists, and public transport [20].

Considering its size, population, and the number of daily commutes within the city, Lille and Vauban-Esquermes become a suitable context for further improvement of and experimentation with the pedestrian mobility. This is even more relevant considering the present development goals of local government [21], which develops existing infrastructure as well new infrastructure to accommodate soft mobility.

2.2 The Catholic University and Junia Camplus

Vauban-Esquermes neighbourhood is also a neighbourhood where the story of Catholic University of Lille begins in the end of the 17th century. First transformation of the area and expansion of university took place in 1873, when due to the lack of space, University decided to reach for new terrains in the newly established quarter Vauban-Esquermes (Fig. 1). An important amount of terrain was bought in the area and first construction projects were developed [22]. The Catholic University of Lille today is composed of 11 institutes and 10 graduate schools, which brings the dynamics to the whole neighbourhood. In this paper, Junia graduate school, which is a part of the Catholic University of Lille, and its construction sites are further analysed [23]. Junia is a higher education institute, joining three graduate schools of engineering HEI, ISA, and ISEN. In September 2019 the YNCREA community (previous name of the Junia institution) presented a real-estate project to be completed by September 2024. The Camplus project must increase the surface available from 29,200 m^2 to 40,300 m^2 with over 22,000 m^2 of new buildings. The main construction works are on the three main campus "Isles": (1) the Maison Legrand, where a new student service area is being created; (2) the Palais Rameau where the revitalisation of an important monument of the city of Lille takes place to expose horticulture to the public; and (3) the Colson buildings which with two new buildings must provide new space for education. The Camplus project has continued despite the pandemic and its completion is expected in December 2025. All this makes Junia an interesting observation site, where the link between the pedestrian mobility and redevelopment allows understanding of interaction between the temporary and permanent residents and the construction project.

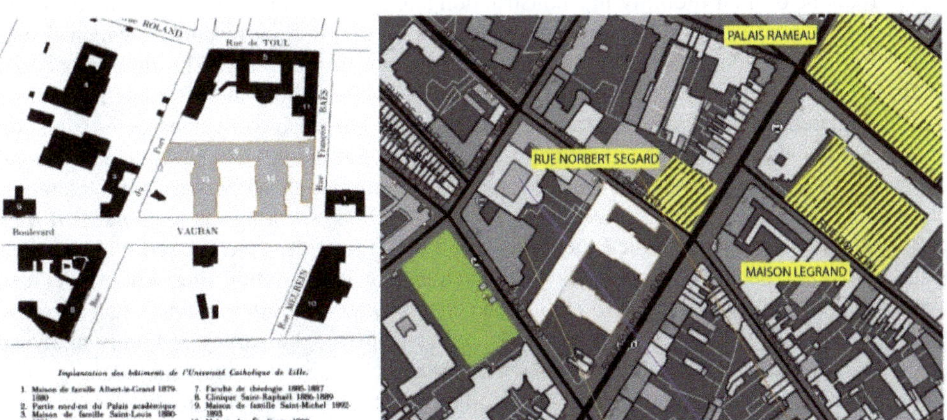

Figure 1: The evolution of the catholic university in time and location of case studies [24]–[26].

3 METHODOLOGY

To understand behaviours of local population in the selected context, proposed methodology is based on three main steps:

- an analysis of the history of Lille's Catholic University and Vauban-Esquermes district using GIS,

- a computer aided web interview (CAWI) survey to study the attitude of pedestrians and the impact of construction sites on their daily routines, and
- observation of three sidewalks in the proximity of Junia construction sites.

The main goal of the research is to analyse the comfort of the sidewalks and pedestrian behaviour, and to propose possible improvements of pedestrian mobility in proximity of construction sites.

3.1 Analysis of the history of Lille's Catholic University and the territory using GIS

The first part of this research was to understand the impact Junia construction site has on pedestrian mobility. The three sites identified are construction sites of the Junia Camplus: two of them are in the Vauban-Esquermes and one is in the Lille Centre neighbourhood. Spatial data is collected from the open data base, including the mapping provided by the metropole of Lille (MEL Open Data). The legal status of how construction sites occupy space is analysed considering "Permisse de stationament", which is a document that must be obtained and exposed to the public before the construction starts [27]. Site visits and photographs of the current situation and real conditions on the site were then collected and added to GIS to allow spatial analysis.

3.2 CAWI survey to study the attitude of pedestrians and their opinion about construction sites

The main purpose of CAWI was to understand the approach of pedestrians, to understand the main problems of their mobility in a construction site districts, and to search for solutions to improve their comfort. The online form consists of 40 conditional responses, which means that interviewees would answer only questions relevant to their case. The first out of four sections of the interview are focused on movement habits, where the respondent provides their living area, working area, the transport method used to reach works, possessed transport vehicles and general rating of the area and why they like or dislike to walk in it. The second section includes questions about the impact on decision making with a series of statements on walking. The third section is specific to construction sites asking the respondent if they pass next to one or more studied locations and the effect of the construction sites on their walking habits. The section concludes with statements on construction sites. The CAWI also allows the possibility of language choice and in this way also international students were included in the study. The survey provided 204 responses. The survey is finally analysed using Python with the use of numerous data processing libraries including Pandas. The analysis includes basic statistical analysis of all numerical variables plotting graphs of all the columns and producing a correlation matrix.

3.3 Observation of three sidewalks located in the proximity of construction sites

One of the main references in this study used for the evaluation of comfort in pedestrian zones is the Pedestrian Comfort Level Guidance for London [8]. This reference is used as it provides a quality method to calculate the comfort of pedestrians on determined sidewalks and give them a grading system from A to E. In this grading system, "A+" stands for a sidewalk in which a pedestrian has a choice to move at whatever speed and in direction they want, while "E" stands for a very limited personal space where movement is very restricted. The assessment suggests measuring flows with the static gate method, "whereby all pedestrians who cross an imaginary line perpendicular to the footway are counted" [17], in specific hours depending on the typology of the road, and 10-minute observation intervals twice an hour to provide the result. To assess the comfort level, video recordings of the pedestrian movement are made during the observation process. The goal of the recording is to observe and photograph common pedestrian behaviours throughout the observation period. The recordings made during the observations are 79 recordings with a total size of over 170 GB. The observation was then analysed, and the counts of the pedestrian flows were counted on a spreadsheet.

4 CASE STUDIES

First, in order to select the case studies an analysis of the types of sidewalks in the Vauban-Esquermes and Lille-Centre districts is made using MEL Open Data, Google Maps Street view, and on-site observations (Fig. 2). A sidewalk typology is then created, and four main types have been identified as follows:

- Local alley: Small alleys including a narrow one-way street, with one pedestrian sidewalk. This type of road is not common in the area (there are only two roads of this type) therefore it is not further observed in this research.
- Local street: A local street confined by buildings with one or more traffic lanes and two pedestrian sidewalks on both sides. There are 16 local streets within the area and an example of such road is the Rue Norbert Segard.
- City boulevard: Large roads confined by buildings consisting of a driveway canalised into more than two lanes. There are four such roads on the site. A case selected for further observations is the Boulevard Vauban in front of the Maison Legrand.
- Green boulevard: Large roads, with large sidewalks on both sides, confined by buildings on the one hand and greeneries on the other hand. A case studied in this research is Boulevard Vauban in front of Palais Rameau.

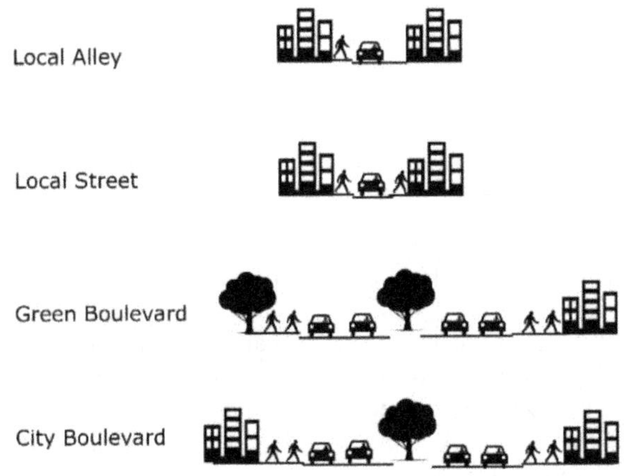

Figure 2: Models of streets design.

For each type of the sidewalk a case study is chosen to be observed and the comfort was assessed according to the Pedestrian Comfort Level Guidance of the city of London [8]. The observations were made in the period between the 22 June and 2 July. Each location was observed for a full weekday. The boulevard sidewalks, were also observed on weekend, following the recommendations of the London Guidance [8]. Video recordings were 10 minutes long and they were taken twice per hour on weekdays and once per hour on weekends. Specific behaviours were photographed along with taking notes to record the flows and pedestrian behaviour and create photograms of the recording. As a result, sheets with typical and atypical behaviours are created and presented assessing the comfort on each location (Fig. 3).

Figure 3: Photographs of behaviours.

4.1 Rue Norbert Segard a case study of a local street

The first case studied is Rue Norbert Segard, a local street often used by students and office workers who every day reach their offices located within the Catholic University of Lille. The road connects the bus stop located on Boulevard Vauban and the metro station "Gambetta" to the entrance of ISA building and the side entrance to the main courtyard of the Lille Catholic University (Fig. 1).

The purpose of the observation in this street is to understand the parking attitudes, the pedestrian behaviour, and to understand what route they choose to reach buildings on the sidewalk next to the construction site. Particular attention is given to pedestrians that are forced to walk on the roadway and directly exposed to vehicles.

The observations of this construction site included counting the flows of people in different directions during a weekday as also observing the people who decide to walk on the road on the side of the construction site to reach the ISA building entrance. The observations were made on a Friday between 7 am and 7 pm with warm weather and light rain in the morning.

The road has been affected by the renovation works of the building named "Maison Legrand" located at the intersection of Boulevard Vauban and Rue Norbert Segard, which causes a total closure of the sidewalk in front of it (Fig. 1). The sidewalk not occupied by the construction works according to the "permisse de stationament" is regularly used as a parking for vehicles and machinery involved in this construction site reducing this way the useful passage from 257 cm to 140 cm, while other machinery and equipment placed next to the cars cause a reduction of the available space to 80 cm.

4.2 Boulevard Vauban in front of Maison Legrand a case study of a city Boulevard

The second case study is on Boulevard Vauban in front of Maison Legrand. It is a city boulevard, which has two large sidewalks separated by trees from the road. Purpose of the observation in this street is to see the modifications of the temporary sidewalks and the pedestrian reaction to obstacles. The observations of this construction site included counting of flows in both directions, observing the attitude of pedestrians who use the temporary sidewalk, and observing the construction workers and the changes in the sidewalk throughout the day. The observations were made on a weekday from 7 am to 7 pm and on a Saturday between 9 am and 5 pm both days of observation were characterised by sunny weather.

The current construction occupies fully the sidewalk in front of it and a temporary sidewalk protected by fences allows pedestrian movement. The sidewalk has irregular sizes as the movement of machinery and parking cause the workers to modify the passage during the day, therefore not allowing a fixed measure.

4.3 Boulevard Vauban in front of Palais Rameau a case study of a green boulevard

The last case study is in Boulevard Vauban in front of the Palais Rameau. The Boulevard is in direct proximity to the park surrounding the Palais Rameau and a college located on Rue Solferino (Fig. 1). The main goal of the observation was to understand the flows of pedestrians and observe their curiosity about the construction site. The observations were made on a weekday from 7 am to 7 pm and on a Saturday between 9 am and 5 pm both days of observation were characterised by sunny weather. The sidewalk has a total width of 380 cm, and one side is separated from the construction site with a fence and panels describing the current constructions and the history of the area. The current construction works do not affect the sidewalk as the plan of the construction uses the area of the park. The mobility of pedestrians, therefore, is not interrupted nor modified with the exception of the park which is currently unavailable.

5 DISCUSSION

5.1 Flows of pedestrian and assessment of comfort level

The observation sessions denoted peak hours on weekdays while flows during weekends were distributed more homogenously on all sites (Fig. 4). The main observations of pedestrian flows and comfort level were as follows:

1. Rue Norbert Segard: the peak of pedestrian flow is between 8 am and 9 am compatible with the office hours and entrance of university students. Second peak hour between 12:00 and 13:30 is compatible with lunchbreak hours, and finally the third peak is at

5 pm, after work. The average flow per hour is of 186 people in both directions, while the peak was of 444 pedestrians per hour. The comfort level assessed is F, assessing very low pedestrian Comfort due to limited width caused by parking of vehicles.

2. Boulevard Vauban/Maison Legrand: the peak flow of pedestrian in front of Maison Legrand was observed at 12 am and an increase in flow was observed after 4:30 pm. The average flow of pedestrians was of 57 pedestrians and the peak flow of 360 pedestrian per hour. The comfort level assessed is F, as the size of the protected pathway does not allow comfortable movement.
3. Boulevard Vauban/Palais Rameau: the peak flow of pedestrian in front of Palais Rameau is more homogenous and observes important peaks after 5 pm. Moreover a few minutes before the beginning of the lessons in the institute located in Rue Solferino and a few minutes after the end of the lessons an important flow of children on bicycles may be noticed. The flow of bicycles is not counted in this study. The comfort level thanks to its large sidewalk has been assessed to be A+. The average flow of pedestrians is of 110 pedestrians per hour and the peak flow is of 240 pedestrians per hour.

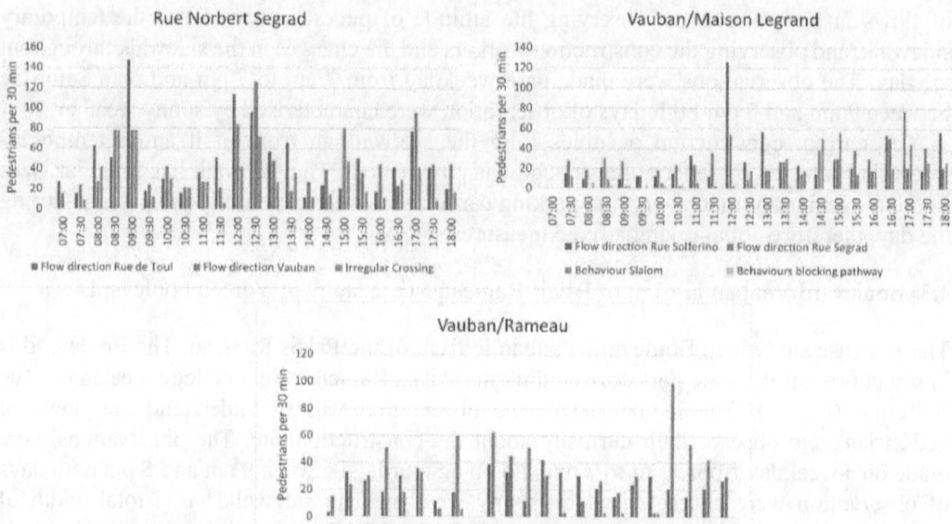

Figure 4: Flow counts during weekday.

5.2 Behaviour observations

The main behaviour observed, especially in the case of the Rue Norbert Segard, was walking on the road and irregular crossing: walking on the road or crossing without an official pedestrian sidewalk accounted 228 occurrences and increases significantly during rush hours (Fig. 3). The second most common type of behaviour, especially in the case of Boulevard Vauban, was a respond to, or a reaction to, physical elements/obstacles (such as cars, machines, and material deposits). The constantly replaced temporary sidewalk (accounting six occurrences), caused confusion of pedestrians who could not find their way. Therefore, an important amount of people (130) decided to take alternative pathways using the road or slaloming between the obstacles deposited on the sidewalk. During the weekend observation the problem was partially solved with a temporary pathway. The third most common behaviour is irregular car parking, with 12 photographed and registered cases. Actors were mainly construction site deliveries and other couriers who dropped their car and left after a few minutes. Another important behaviour is the curiosity towards the construction sites and to its history reflected actions such as peeking through the panels, reading information, looking through the fence, etc. The observation allowed understanding the frequency of reaction to panels in both the Maison Legrand construction and the Palais Rameau, showing more interest during weekends. The panels located in front of Palais Rameau were partially covered with graffiti, which did not discourage pedestrians to look at the panels, with seven people reading panels contents on weekday and over 10 on weekend. Another important behaviour includes construction workers who redirect traffic and use the road without personal protective equipment (PPE) required and recommended for road operations. At last, a dangerous behaviour often repeated by all road users was jaywalking of main roads observed within eight photograms of people deciding to neglect the designed crossing on road intersections.

5.3 Survey results

The CAWI was conducted through Microsoft forms software and was distributed through a QR code through posters and leaflets distributed in the proximity of the Catholic University of Lille. The survey was also distributed electronically with the usage of the university mailing lists. The total amount of answers was 203 of which 194 of people living or working in the MEL. The period of distribution started on the first of June 2022 and was terminated on 11 July 2022. 33% of participants affirmed that their main transport method is walking, followed by 29% by car and 24% by public transport. Main reasons for the decision of their dominant transport method are simplicity, over 60% of respondents, and time, over 50% of respondents. The respondents were asked to rate the walkability of the area answering with an average rating of 3.66, with more than 85% of respondents generally satisfied of the area rating it level three or higher. When it comes to the statements about walkability and construction sites, almost 32% of respondents had a strongly negative opinion on the parking of cars compatible with the observations previously made, and over 41% of them did not feel recognised by construction companies. Most pedestrians accept issues related to construction if they provide improvements to the area, and surprisingly they are comfortable walking in their proximity (under scaffolding, for example). Respondents feel safe on sidewalks and find pleasure in walking, especially when in proximity of green spaces. Correlation of data, with the use of correlation provided by Pandas function Pearson correlation function, allowed finding an important correlation between travel time and transport method (Fig. 5). The noticeable fact is that users who choose walking as their main transport method generally have the lowest commute time, while the highest times is for those using public transport.

6 CONCLUSION

The current study shows that pedestrian mobility in proximity of construction sites triggers different, often unpredictable, and unsafe behaviours. However, it is obvious from the survey and observations, that respondents are curious about and feeling comfortable in a proximity of construction site. Therefore, walkable spaces in proximity of construction site may be produced with the use of construction elements and machinery, which would allow an extension of sidewalks inside construction sites providing a safe passage for pedestrians. The boundary between construction sites and pedestrian zones would blur, and communication between users on both sides would improve.

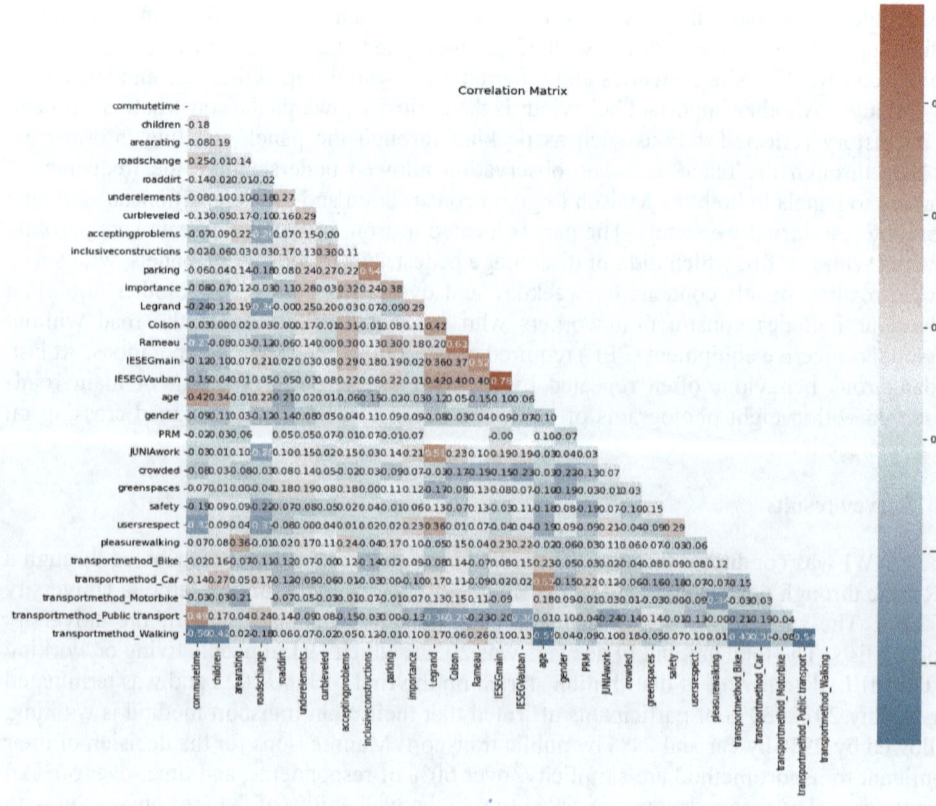

Figure 5: Correlation matrix of answers to CAWI interview.

As one solution, for the city boulevard sidewalks a more detailed and real-time plan of protected pathways could be established for both pedestrians and workers, who would then more easily apply to these solutions avoiding unsafe behaviours. The plan of pedestrian mobility addressed by the "permisse de stationament" and "permisse a voire" could consider a wider area affected by the activities on the site, especially considering daily traffic. This could be achieved with a parking plan that includes both permanent and temporary vehicles and machines used on the site. The legislative process for obtaining such a permission could include a plan of pedestrian mobility to compensate the issues caused by the occupation of sidewalks. Moreover, by avoiding narrow and uncomfortable sidewalks, a comfortable mobility through the city which encourages pedestrian mobility would be achieved.

This is especially relevant in compact and dense urban areas like Lille, where the proximity of services and its existing infrastructure allow comfortable movement to its permanent and temporary residents. In that sense, it is possible to address Road safety Policies through small urban actions and to address current climate crisis at local scale.

ACKNOWLEDGEMENT
We would like to thank the Smart and Resilient Cities research team of Junia – HEI for advising, support, and providing funding and equipment for this research.

REFERENCES

[1] Balaban, O., The negative effects of construction boom on urban planning and environment in Turkey: Unraveling the role of the public sector. *Habitat International*, **36**, pp. 26–35, 2012. DOI: 10.1016/j.habitatint.2011.05.003.

[2] Ng, C.F., Effects of building construction noise on residents: A quasi-experiment. *Journal of Environmental Psychology*, **20**, pp. 375–385, 2000. DOI: 10.1006/jevp.2000.0177.

[3] Hong, J., Kang, H., Hong, T., Park, H.S. & Lee, D.E., Construction noise rating based on legal and health impacts. *Automation in Construction*, **134**, 2022. DOI: 10.1016/j.autcon.2021.104053.

[4] Hong, J. et al., An empirical analysis of environmental pollutants on building c4onstruction sites for determining the real-time monitoring indices. *Building and Environment*, **170**, 106636, 2020. DOI: 10.1016/j.buildenv.2019.106636.

[5] Çelik, T., Arayici, Y. & Budayan, C., Assessing the social cost of housing projects on the built environment: Analysis and monetization of the adverse impacts incurred on the neighbouring communities. *Environmental Impact Assessment Review*, **77**, pp. 1–10, 2019. DOI: 10.1016/j.eiar.2019.03.001.

[6] Consolidated text: Regulation (EC) No 1221/2009 of the European Parliament and of the Council of 25 November 2009 on the voluntary participation by organisations in a Community eco-management and audit scheme (EMAS), repealing Regulation (EC) No 761/2001 and Commission Decisions 2001/681/EC and 2006/193/EC.

[7] Janicak, C.A., Pedestrian fatalities in the construction industry. *Professional Safety*, **46**, pp. 16–19, 2001.

[8] Finch, E., *Pedestrian Comfort Level Guidance*, 2nd ed., Transport for London: London, 2022. https://content.tfl.gov.uk/pedestrian-comfort-guidance-technical-guide.pdf. Accessed on: 8 Sep. 2022.

[9] Strohmeier, F., Barriers and their influence on the mobility behavior of elder pedestrians in urban areas: Challenges and best practice for walkability in the city of Vienna. *Transportation Research Procedia*, **14**, pp. 1134–1143, 2016. DOI: 10.1016/j.trpro.2016.05.184.

[10] Carlson, J.A. et al., Impacts of temporary pedestrian streetscape improvements on pedestrian and vehicle activity and community perceptions. *Journal of Transport and Health*, **15**, 2019. DOI: 10.1016/j.jth.2019.100791.

[11] General Assembly Resolution 71/313, Work of the Statistical Commission pertaining to the 2030 Agenda for Sustainable Development, A/RES/71/313 (10 July 2017). https://undocs.org/en/A/RES/71/313.

[12] Communication from the Commission to the European Parliament, the Council, the European Economic and Social Committee and the Committee of the Regions, Towards a European road safety area: Policy orientations on road safety 2011–2020.

[13] Santilli, D., D'Apuzzo, M., Evangelisti, A. & Nicolosi, V., Towards sustainability: New tools for planning urban pedestrian mobility. *Sustainability*, **13**, 2021. DOI: 10.3390/su13169371.

[14] Shaw, J., Chitturi, M., Han, Y., Bremer, W. & Noyce, D., Bicyclist and pedestrian safety in work zones: Recent advances and future directions. 2016.

[15] Chiffres clés, 2022. https://www.lillemetropole.fr/votre-metropole/institution/territoire-de-la-mel/chiffres-cles. Accessed on: 9 Sep. 2022.

[16] Lille.fr, Construire ensemble un quartier durable et aménager les friches, 2022. https://www.lille.fr/Vauban-Esquermes/Actualites/Construire-ensemble-un-quartier-durable-et-amenager-les-friches. Accessed on: 8 Sep. 2022.

[17] INSEE, Population en 2018 Recensement de la population: Base infracommunale (IRIS). 2021.
[18] Office de Tourisme et des Congrès de Lille, Histoire de Lille, 2022. https://www.lilletourism.com/histoire-de-lille.html. Accessed on: 8 Sep. 2022.
[19] S.V.A.H. Ville de Lille, Focus La Citadelle de Lille, 2020. https://parcdelacitadelle.lille.fr/node/486/deux-publications-autour-de-la-citadelle. Accessed on: 8 Sep. 2022.
[20] Lille.fr, À Lille, déplacez-vous autrement! 2022. https://www.lille.fr/www_en/Actualites/A-Lille-deplacez-vous-autrement-!. Accessed on: 8 Sep. 2022.
[21] Aménagement d'espaces partagés, 2022. https://www.lillemetropole.fr/votre-metropole/competences/cadre-de-vie/amenagement-despaces-partages. Accessed on: 23 Sep. 2022.
[22] Ribaut, J.P., La construction des bâtiments de l'université catholique de Lille. *Revue d'histoire de l'Église de France*, pp. 77–85, 1987.
[23] Université Catholique de Lille, Nos établissements et institutions, 2022. https://www.univ-catholille.fr/nos-etablissements-et-institutions. Accessed on: 8 Sep. 2022.
[24] Google Maps, Université Catholique de Lille. http://maps.google.com.
[25] OpenData MEL, Explore, 2022. https://opendata.lillemetropole.fr/explore/?sort=modified. Accessed on: 9 Sep. 2022.
[26] Cadastre Napoleonien, Recherche, 2022. https://archivesdepartementales.lenord.fr/search/results?formUuid=b57520fe-2768-4335-b138-119f274f0a73&sort=relevance&mode=list&facet_media=image&0-controlledAccessGeographicName%5B%5D=LILLE. Accessed on: 9 Sep. 2022.
[27] Démarches l'Intérieur, Ministère de l'Intérieur, 2022. https://www.demarches.interieur.gouv.fr/professionnels/permis-stationnement-permission-voirie. Accessed on: 8 Sep. 2022.

RISK-BASED TUNNEL DESIGN FOR CONSEQUENCES OF ROAD ACCIDENTS: THE ROLE OF TUNNEL LENGTH

ANTONELLA PIREDDU[1], MARA LOMBARDI[2], SILVIA BRUZZONE[3] & DAVIDE BERARDI[2]
[1]Italian National Institute for Insurance against Accidents at Work (INAIL), Italy
[2]Sapienza Università di Roma, Italy
[3]Italian National Institute of Statistics (Istat), Italy

ABSTRACT

Tunnel extension is an under-analysed variable in road tunnel accidents despite being a dimensioning parameter for the purposes of users' safety according to Directive 2004/54/EC. Recent studies have shown a correlation between the tunnel length and consequences of accidents. The analysis of fire events which occurred in tunnels indicates that in many cases fires are triggered by road accidents. By analysing the road accidents in Italy, the study aims to assess the relative risk of accidents with serious consequences for different classes of road tunnels. The second objective was to assess, using a vehicle type (or size) approach, the corresponding probability of accidents involving vehicles or trucks and special vehicles resulting in serious consequences (domino effect). We analysed the Italian National Institute of Statistics (Istat) dataset on tunnel accidents which occurred between 2018 and 2020 on Italian public roads, involving at least one vehicle. Of these, we extracted tunnel accidents, classified by tunnel length and estimated the corresponding probability of serious consequences. The analysis identified 1,885 case studies of tunnel accidents that occurred in approximately 265 long tunnels and 450 short tunnels and underpasses. Compared with "controls", "size" was found to be more than double in long tunnels where the related probability of serious accident consequences exceeded 50% more than those of short tunnels. We found that the related probability associated with serious accident consequences in tunnels over 500 m in length was higher than in short tunnels, except for trucks and special vehicles. Road accidents and research on risk evaluation of the effects associated with long and short tunnels are rare. The study aims to fill these gaps.

Keywords: road tunnel safety, tunnel length, vehicle, risk exposition, work-related road risk, risk based design.

1 INTRODUCTION

Studies are not consistent in regards to the impact of tunnel length on safety indicators. However the length effect was indirectly considered in Directive 2004/54/CE, as a parameter for defining classes of minimum safety requirements. A road accident occurs when two or more vehicles collide or when a vehicle hits a pedestrian, an animal or an object such as a tunnel wall, traffic sign, etc. Amundsen [1], carried out a series of tunnel traffic accident analyses on Norwegian road tunnels from 1994 to 2009. They observed that the risk of accidents in tunnels was lower than on open roads, but that the severity of accidents was much higher (Amundsen [1], Amundsen and Engebretsen [2] and Amundsen and Ranes [3]). In general, it was found that the frequency of tunnel accidents is lower than on straight sections, curves, roundabouts and intersections. Tunnel extension is an under-analysed variable in road tunnel accidents although being a dimensioning parameter for the purposes of users' safety according to Directive 2004/54/EC [4]. Recent studies have found a positive correlation with the length and consequences of tunnel accidents (Caliendo et al. [5]), other studies have shown a higher absolute and relative frequency of road accidents and injuries in the shortest tunnels (Pireddu and Bruzzone [6]).

However, such studies have not yet demonstrated a clear correlation between the risk of accident and the length of the fornix tunnel. This is probably the result of the unavailability of sufficiently comprehensive case studies to draw reliable conclusions. According to

WIT Transactions on The Built Environment, Vol 214, © 2022 WIT Press
www.witpress.com, ISSN 1743-3509 (on-line)
doi:10.2495/SSR220061

previous simulation and psychological studies, the analytic results indicate that the different types of tunnel have distinct accident characteristics and therefore should be considered separately for safety analysis (Pervez et al. [7], PIARC [8]).

Based on recent findings, fire in tunnels are in many cases caused by road accidents that do not occur in the same conditions as the open roads. Although less frequent than in open sections, road accidents in tunnels can cause serious fires. A notorious example is the fire in the Gotthard Tunnel on 24 October 2001 where, at the entrance to the tunnel, the driver of a truck carrying tyres lost control of the vehicle, invading the other lane and hitting a truck that ran through the tunnel in the opposite direction. The frontal collision between the two vehicles caused the fire, whose flames extended for about 300 m, raising the temperature to 1,200°C. This event highlights the importance of empirical analyses of road accidents and related variables in the accident scenario. The information derived from case studies is a useful tool in road safety design to identify appropriate measures (or safety minimum requirements, according to Directive 2004/54/CE) to prevent accidents in tunnels. Consequently, these analyses are useful in assessing the risk of road accidents and the risk of fire triggered by a domino effect, in road tunnels.

Therefore, the main purpose of this study, based on road accidents in Italian tunnels, was to assess the relative risk of road accidents resulting in serious consequences for road users and workers, in various types of road tunnels. A second objective was to assess, using a vehicle type approach, the relative risk of accidents resulting in serious consequences, in terms of deaths or serious injuries, when vehicles or truck and special vehicles are involved.

2 MATERIAL AND METHODS

Case studies on road accidents are provided by Istat [9]–[13], based on the "Survey on road accidents resulting in death or injury" that includes all road accidents involving deaths within 30 days or injuries. This archives contain accidents occurred throughout the country in public roads where "at least one vehicle is involved and where at least one injured person is recorded by a police authority" (Vienna Convention on Road Traffic, 1968). The detection data refers to the time which the accident occurred (Istat data warehouse, I. Stat, 2018–2020).

The Istat dataset, provided the following information: the accident circumstances (European Commission [14]), the accident type, the carriageway, the consequences (fatalities or injuries), the involvement of pedestrian, the geographical coordinates (Cima et al. [15], Costabile et al. [16]), the road type, the time of the accident, the journey purpose (work-related or not work-related) [17], the vehicle type. By geo-processing, we integrated the Istat dataset with the OpenStreetMap (OSM) road information to obtain the tunnel section geometry involved in the accidents. According to Directive 2004/54/EC [4], tunnels were grouped into two length classes: up to 500 m (underpasses included), over 500 m, more suitable for our investigation (Fig. 1 and Table 1) because equipped with minimum requirements according to the tunnel length. The tunnels involved in at the least one accident in 2018–2020 were then ordered by frequencies, accident variable and classes. This first step of study missed variables such as the characteristics of road sections, the tunnel route and traffic intensity-composition. We implemented a case-control design for the estimation of the association between the tunnel length and severity of road accident consequences. We compared population exposure for two clusters: accidents resulting in serious consequences occurred in long tunnels and remaining accidents without consequences (control cases). The dichotomous dependent variable Y assumed 0 value in case of non-serious accident consequences and 1 in case of serious consequences. Among the risk factors (regressors Xi, i = 1 ... n in the n·m matrix), we have taken the ones affecting the response of the variable Y. Then, by means of logistic regression (Breiman et al. [18]) and R applications, we estimated the incidence risk ratio. The relation between the severity of consequences Y and the other independent variables Xi is explained by eqn (1):

$$= \ln \frac{p}{(1-p)} = \beta 0 + \beta 1 \cdot X1 + \beta 2 \cdot X2 + \ldots \beta n \cdot Xn + \varepsilon. \quad (1)$$

Probability of serious consequences Y is provided by eqn (2). The coefficients βi of exponentials associated with independent variables Xi, represent the event occurrence odds ratio (OR) corresponding to the increment of the independent variable (e.g. the length), net of other ones

$$p = \frac{1}{e^{-(\beta 0 + \beta 1 \cdot X1 + \beta 2 \cdot X2 + \ldots \beta n \cdot Xn + \varepsilon)}} + \varepsilon. \quad (2)$$

The exp (βi) provided a measure of the relative risk (probability of serious consequence in case of road accident) Yi, compared to the exposure to the risk factor Xi, represented by the length of the tunnel over 500 m. We implemented eqns (1) and (2), the "glm" function and the "family = binomial" to the following subsets including: (i) accidents involving all vehicle type; (ii) accidents involving all vehicle type excluded trucks or special vehicles; (iii) accidents involving at least one truck or special vehicle. As a result, we obtained a cross validated model that estimates, for each subset, the "size" of severity of accident consequences by length classes. Within the subsets (ii) and (iii), using kernel density, which is a non-parametric method, we estimated the density and the probability that a certain class of accident consequences belonged to a given class of tunnel length (Fig. 2).

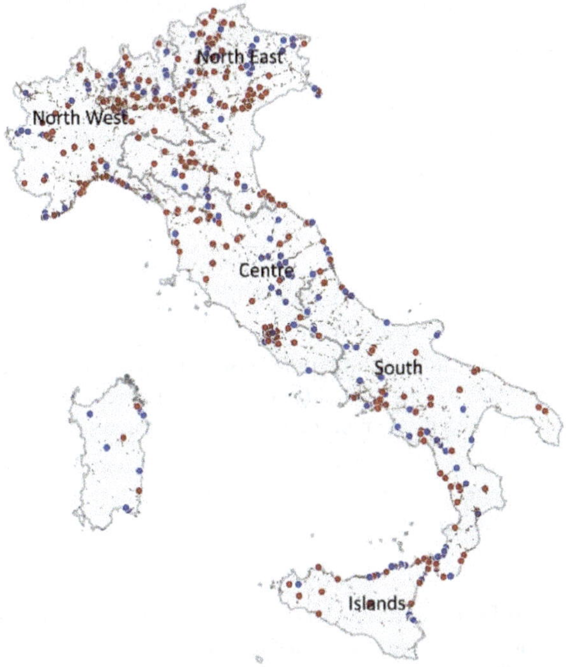

Figure 1: Location of the studied tunnels. Long tunnel accidents (blue) and short tunnel accidents (red), Italy, 2018–2020. *(Source: Author's processing from Istat and OSM dataset. QGIS.)*

Table 1: Attributes and labels resulting from the Istat data, Italy, 2018–2020. *(Source: Author's processing from Istat data, QGIS and R Studio.)*

Variables (attribute)	Classes (label)
Accident circumstances	distance (not keeping distance between vehicles); distraction; normal driving; speeding; other_circ (unspecified and other circumstances); corresponding to driver behaviour recorded when the accident occurred (Amundsen [1], Amundsen and Engebretsen [2], Pireddu and Bruzzone [6], Gariazzo et al. [19], [20])
Accident type	rear-end; collision, impact (impact with other vehicles, frontal or lateral collisions); other (pedestrians or obstacles, skidding or off-road, etc.)
Carriageway	carr1 (one-way lane); carr2 (two-way lanes); carr3 (two carriageways); carr4 (more than two carriageways)
Consequence	serious (accidents with more than three injuries and/or one or more fatalities); not-serious (accidents with up to three injuries and no fatalities)
Journey purpose	work-related 1 (driving for duty); work-related 2 (as part of commuting); not work-related (journey purpose not work-related) (Pireddu and Bruzzone [6], [21]) The journey purpose resulted often not filled due to the difficulty of recording this information at the scene of the accident
Pedestrian	0 (accident not involving pedestrian); 1 (accident involving at least one pedestrian)
Road tunnel location (road type)	motorway (road inside or outside urban areas, reserved to certain categories of vehicles); rural (road outside urban areas and not motorway); urban (road in urban areas and not motorway) (Italian Road Traffic Law [22])
Time of accident	morning (06:00–12:00); afternoon (13:00–18:00); evening (19:00–21:00); night (22:00–05:00); defined according to a conventional interval
Tunnel length (class)	short tunnels and underpasses; (up to 500 m in length or "0–500"); long tunnel (over 500 m in length or ">500"); unclassified (when the tunnel location were not available)
Vehicle type	car; truck or special vehicle (heavy goods vehicle); motorcycle; otherV (other vehicle); bicycle, scooter; (bicycle, electric bicycle and scooter)

3 RESULTS

The 1,885 cases selected from the Italian road accidents dataset, complete with information on the accident scenario, tunnel geometry and traffic information, included 2,999 injured people and 60 deaths. The accident dataset provided about 715 tunnels (Fig. 1) exactly classified: 37% long tunnels (>500 m) and 63% short tunnels (0–500 m).

Table 1 shows a classification of accident variables, used in the $n \cdot m$ matrix on accident frequencies where the rows are the tunnels studied and columns the classes.

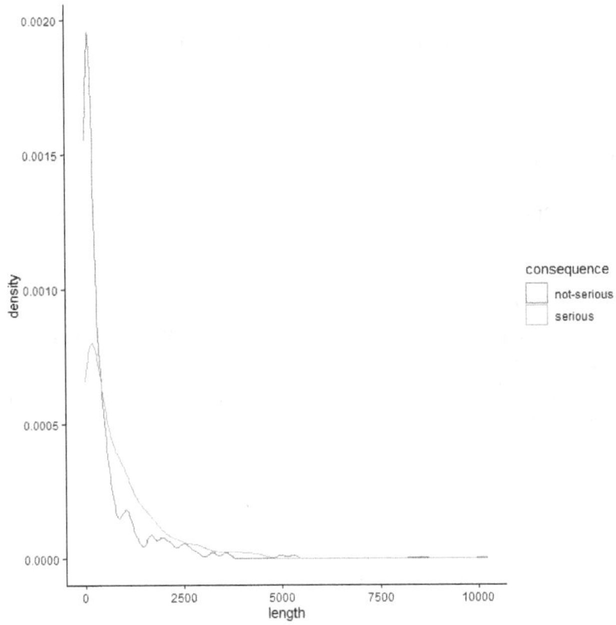

(a)

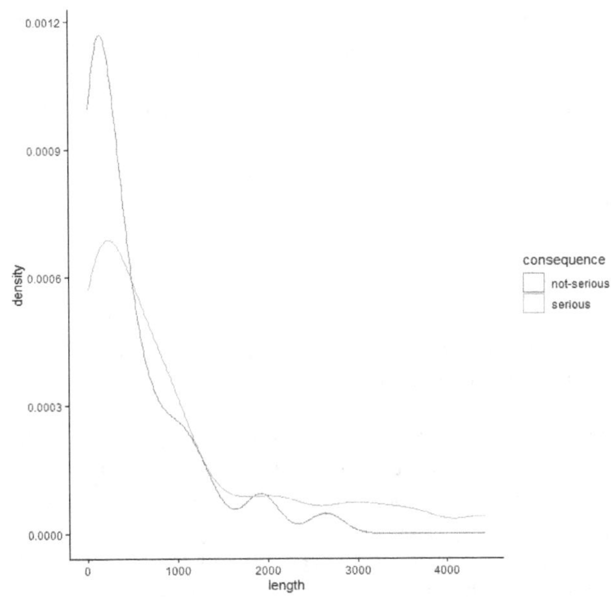

(b)

Figure 2: Kernel density estimation. Consequences of serious (turquoise) and not serious (red) accidents by tunnel length (m). (a) Accidents including all vehicle type trucks and special vehicles excluded; and (b) Accidents involving at least one truck or special vehicle. Italy, 2018–2020. *(Source: Author's elaboration from Istat data.)*

3.1 Association between accident consequences and tunnel length by vehicle class

Applying eqns (1) and (2), the "glm" function and the "family=binomial" in R to the three subsets (i), (ii), (iii), we obtain the probability of serious consequences associated to long tunnel exposure or relative risk (Tables 2 and 3).

3.2 The risk ratio and the relative risk attributable to exposure to tunnels longer than 500 m

The incidence relative risk ratio allows comparison of accident rates between two different groups. Attributable relative risk is the increase or decrease in the probability of serious consequences (outcome) that is attributable to road users' exposure in long tunnels (Table 3). This parameter provides a measure of the absolute frequency of the outcome associated with exposure. Considering the outcomes per 100 population units, Table 3 lists the confidence intervals.

As reported in Table 2, the odds reached values of 1.7, 1.9 and 2.9 for subsets (i), (ii) and (iii). The odds ratio reached for the same subsets 2.69, 2.70 and 2.45 while the incidence relative risk ratio 1.58, 1.62 and 1.37 respectively, with a 95% confidence interval. For the subset (i), the incidence relative risk ratio (RR) falls within a confidence interval of 1.37 ÷ 1.82 and thus is always greater than 1, indicating that the incidence rate is higher in the group exposed to long tunnels (+) than in the control group (−). The result indicates that for vehicles as a whole (i), exposure to long tunnels is a factor that increases the probability of serious accident consequences by 58%. This result is even higher in the subset where no trucks or special vehicles are involved, where exposure to long tunnels leads to a 62% increase in relative risk (C.I. 1.38 ÷ 1.90). On the other hand, in the case where each accident involves at least 1 among trucks or special vehicles, the incidence relative risk ratio is lower than the previous ones (1.37) and falls in the confidence interval between 1.04 and 1.79. This means that the incidence relative risk ratio can reach as high as 1, thus indicating that the severity of accident consequences in long tunnels can be very close to that observed in short tunnels. For subset (iii), unlike subsets (i) and (ii), we have accepted RR = 1 and the null hypothesis associated with the probability that the severity of accident consequences in the long tunnels coincides with that observed in the short tunnels.

3.3 The consequences of road tunnel accidents referring to the length of the tunnel and the type of vehicle involved

Not serious consequences (up to three injuries in the 3-year period) affected 354 tunnels out of the total, while serious consequences (more than three injuries and/or at least one death in three years) affected 361 tunnels (64% long tunnels and 36% short tunnels).

The Kernel graphs in Fig. 2 present density curves by tunnel length expressed in metres (x-axes) for severe and not serious consequences. When heavy vehicles were not involved in accidents (ii), the density graph reaches 0.0020 for not serious consequences and 0.0008 for serious consequences. The curve reverses between 500 m and 5,000 m, when severe consequences exceed not serious consequences (Fig. 2(a)).

If heavy vehicles are involved in accidents (iii), the density reaches 0.00012 for non-serious consequences and 0.0008 for serious consequences. The curve reverses between 500 m and about 1,000 m in length, where serious consequences exceed not serious consequences (Fig. 2(b)).

Table 2: The accident outcomes. Long tunnels exposure. Probability and relative risk associated to accidents involving (i) all type of vehicles; (ii) trucks or special vehicles excluded; and (iii); trucks or special vehicles. Italy, 2018–2020. *(Source: Author's processing from Istat and OSM data on road accidents, QGIS and R Studio.)*

Tunnel type and consequence	The accident outcomes for accidents involving all type of vehicles (i)				
	Serious consequences (Outcome +)	Not serious consequences (Outcome −)	Total consequences	Incidence probability	Odds
>500 m (exposed +)	174	91	265	65.7	1.9
≤500 m (exposed −)	187	263	450	41.6	0.7
Total	361	354	715	50.5	1.0

Tunnel type and consequence	The accident outcomes for accidents involving all type of vehicles, trucks or special vehicles excluded (ii)				
	Serious consequences (Outcome +)	Not serious consequences (Outcome −)	Total consequences	Incidence probability	Odds
>500 m (exposed +)	136	75	214	63.6	1.7
≤500 m (exposed −)	150	232	382	39.3	0.6
Total	286	310	596	48.0	0.9

Tunnel type and consequence	The accident outcomes for accidents where trucks or special vehicles are involved (iii)				
	Serious consequences (Outcome +)	Not serious consequences (Outcome −)	Total consequences	Incidence probability	Odds
>500 m (exposed +)	38	13	51	74.5	2.9
≤500 m (exposed −)	37	31	68	54.4	1.2
Total	45	74	119	63.0	1.7

Table 3: The accident consequences. Exposure to long tunnels. Point estimates for accidents involving (i) all type of vehicles; (ii) trucks or special vehicles excluded; and (iii) accidents where trucks or special vehicles are involved (95% CI). Italy, 2018–2020. *(Source: Author's processing from Istat and OSM data on road accidents, QGIS and R Studio.)*

Accidents involving all type of vehicles (i)	Point estimates	Confidence interval 95%
Incidence relative risk ratio	1.58	(1.37 ÷ 1.82)
Odds ratio	2.69	(1.96 ÷ 3.69)
Attributable relative risk in the exposed	24.10	(16.80 ÷ 31.41)
Attributable fraction in the exposed (%)	36.71	(27.20 ÷ 44.98)
Attributable relative risk in the population	8.93	(3.09 ÷ 14.78)
Attributable fraction in the population (%)	17.69	(11.77 ÷ 23.22)

For subset (i), uncorrected $\chi 2$ test that OR = 1: $\chi 2$ (1) = 38.767. Pr> $\chi 2$ = <0.001. Fisher exact test that OR = 1: Pr> $\chi 2$ = <0.001. Wald confidence limits.

Accidents involving all type of vehicles trucks or special vehicles excluded (ii)	Point estimates	Confidence interval 95%
Incidence relative risk ratio	1.62	(1.38 ÷ 1.90)
Odds ratio	2.70	(1.91 ÷ 3.81)
Attributable relative risk in the exposed	24.28	(16.19 ÷ 32.38)
Attributable fraction in the exposed (%)	38.21	(27.44 ÷ 47.39)
Attributable relative risk in the population	8.72	(2.39 ÷ 15.05)
Attributable fraction in the population (%)	18.17	(11.45 ÷ 24.38)

For subset (ii), uncorrected $\chi 2$ test that OR = 1: $\chi 2$ (1) = 32.408. Pr> $\chi 2$ = <0.001. Fisher exact test that OR = 1: Pr> $\chi 2$ = <0.001. Wald confidence limits.

Accidents where trucks or special vehicles are involved (iii)	Point estimates	Confidence interval 95%
Incidence relative risk ratio	1.37	(1.04 ÷ 1.79)
Odds ratio	2.45	(1.11 ÷ 5.40)
Attributable relative risk in the exposed	20.10	(3.27 ÷ 36.93)
Attributable fraction in the exposed (%)	26.97	(4.30 ÷ 44.27)
Attributable relative risk in the population	8.61	(−6.06 ÷ 23.29)
Attributable fraction in the population (%)	13.67	(0.7 ÷ 24.94)

For subset (iii), uncorrected $\chi 2$ test that OR = 1: $\chi 2$ (1) = 5.051 Pr> $\chi 2$ = 0.025. Fisher exact test that OR = 1: Pr> $\chi 2$= 0.034. Wald confidence limits.

4 DISCUSSION

The purpose of this case-control study was to investigate the relationship between the consequences of traffic accidents and the extent of tunnels. The objective was also to analyse the same phenomenon with respect to different types of vehicles involved. The three accident subsets considered at this purpose included (i) the entirety of vehicles involved in the original dataset; (ii) the previous set, trucks and special vehicles excluded; and (iii); and the subset of accidents where at least one truck or special vehicle was involved.

Amundsen [1], Amundsen and Engebretsen [2] and Amundsen and Ranes [3] observed that the relative risk of accidents in tunnels was lower than on open roads, but that the severity of accidents was much higher. Accident severity is higher when the vehicle collides against the tunnel wall than when it collides against the guardrail, on open roads. This is compounded by the fact that there is reduced accessibility of rescue devices such as cranes in tunnels (Lemke [23]). Caliendo et al. [5] and Lemke [23] found that tunnel length negatively affects road safety because drivers tend to demonstrate less concentration in long tunnels. In addition, it has been shown through a negative binomial regression model for non-serious and serious accidents that the frequency of accidents on unidirectional tunnel increases with tunnel length, in addition to other factors (Caliendo et al. [5]). The result of our study referred overall to motorway, rural and urban roads, showed a correlation between the severity of consequences and tunnel length.

Bassan [24] included an overview of traffic safety and design aspects in road tunnels and, in addition, discussed the severity of accidents in road tunnels, including fires and their relationship to road accidents. Nævestad and Meyer [25] analysed factors associated with vehicle fires and smoke without fire (SWF) in Norwegian road tunnels. On 2008–2011, they found that an average of 21% of vehicle fires were caused by a road accident. In other words, the accident itself may be the direct cause of the fire. They also found that technical problems were the most frequent cause of fires and SWF accidents in heavy vehicles, while single-vehicle and multi-vehicle accidents were the most frequent cause of fires in vehicles weighing less than 3.5 tons (Nævestad and Meyer [25]). The largest contributor to relative risk in road tunnels is collisions and other types of road accidents (Table 1). Fires, engine or brake failures are also events that must be considered in the risk assessment of road tunnels. Similarly for rare events with potentially important consequences, such as the transport of dangerous goods (Bassan [24]).

Since the vehicle type most involved in accidents is the car, some deviation may affect the overall interpretation of the phenomenon studied. This leads to non-exhaustive conclusions about the prevalence of severe outcomes in truck or special vehicle populations. According to Xing et al. [26], vehicle type was found to be a determinant of the severity of individual tunnel accidents. Compared with cars, trucks and special vehicles had a significantly lower risk of accidents. This result can be explained by the difference in structure and speed between heavy vehicles and cars (Xing et al. [26]). Our analysis showed that the risk of serious consequences in long tunnels is higher for subset (i) (RR: 1.58; 95%; CI: 1.37 ÷ 1.82) and (ii) (RR: 1.62; 95%; CI: 1.38 ÷ 1.90). Vice versa, when considering the subset of accidents where at least one truck or the special vehicles was involved, the risk ratio (RR: 1.37) falls within a 95% confidence interval close to 1 (1.04 ÷ 1.79). Therefore, when trucks and special vehicles are involved in accidents (iii) the probability of serious accidents in long tunnels may be assumed to be the same as in short tunnels.

5 CONCLUSIONS

This case-control study was applied to real-world tunnel road accident data comprehensive in location and tunnel length involved, occurred in Italy over 2018–2020. The case-control design point out an association, to be further analysed, between the tunnel length and the consequences of road accident with injured and fatalities.

We found tunnel over 500 m, to be positively associated with serious consequences if not heavy vehicles resulted involved in the accident scenario. If heavy vehicles were involved the incremental risks achieved 38% with a CI close to 1. In this case the probability of serious accident consequences can be assumed equal than the short tunnels.

The methodology proposed in our study as well as the results obtained represent a useful tool in the risk analysis that precedes risk assessment and from which risk prevention measures are derived.

REFERENCES

[1] Amundsen, F.H., Studies of driver behaviour in Norwegian road tunnels. *Tunnelling and Underground Space Technology*, **9**(1), pp. 9–15, 1994.

[2] Amundsen, F.H. & Engebretsen, A., Studies on Norwegian road tunnels II. An analysis on traffic accidents in road tunnels 2001–2006, 2009.

[3] Amundsen, F.H. & Ranes, G., Studies on traffic accidents in Norwegian road tunnels. *Tunnelling and Underground Space Technology*, **15**(1), pp. 3–11, 2000.

[4] European Parliament, Directive 2004/54/EC of the European Parliament and of the Council of 29 April 2004 on minimum safety requirements for tunnels in the Trans-European Road Network, 2004.

[5] Caliendo, C., De Guglielmo, M.L. & Russo, I., Analysis of crash frequency in motorway tunnels based on a correlated random-parameters approach. *Tunnelling and Underground Space Technology*, **85**, pp. 243–251, 2019.

[6] Pireddu, A. & Bruzzone, S., An analysis of the influence of tunnel length and road type on road accident variables. *Rivista di Statistica Ufficiale*, **2**(2021), pp. 71–102, 2021. https://www.istat.it/it/ archivio/262320.

[7] Pervez, A., Huang, H., Han, C., Wang, J. & Li, Y., Revisiting freeway single tunnel crash characteristics analysis: A six-zone analytic approach. *Accident Analysis and Prevention*, **142**, 2020. DOI: 10.1016/j.aap.2020.105542.

[8] PIARC, Comité technique 3.3: Exploitation des tunnels routiers. Human factors and road tunnel safety regarding users. Paris, La Défense CEDEX, PIARC: France, 2008.

[9] Istituto Nazionale di Statistica (Istat) and Automobile Club d'Italia (ACI), Incidenti stradali in Italia. Comunicato Stampa, 2018. https://www.istat.it/it/archivio/232366.

[10] Istituto Nazionale di Statistica (Istat), Rilevazione degli incidenti stradali con lesioni a persone. Informazioni sulla Rilevazione. https://www.istat.it/it/archivio/4609.

[11] Istituto Nazionale di Statistica (Istat), Survey on road accidents resulting in death or injury. *SIQual – Information System on Quality of Statistical Production Processes.* http://siqual.istat.it/SIQual/lang.do?language=UK.

[12] Istituto Nazionale di Statistica (Istat), Your direct access to the Italian Statistics: The complete data warehouse for experts. http://dati.istat.it/Index.aspx.

[13] Istituto Nazionale di Statistica (Istat) and Automobile Club d'Italia (ACI), Incidenti stradali in Italia. Comunicato Stampa, 2020. https://www.istat.it/en/archivio/259830.

[14] European Commission, Community Database on Road Accidents – CARE, 2016. Road Safety: New statistics call for fresh efforts to save lives on EU roads, Press Release, European Commission: Brussels, 31 March 2016.

[15] Cima, V., Carroccio, M. & Maseroli, R., Corretto utilizzo dei Sistemi Geodetici di Riferimento all'interno dei software GIS. *Proceedings of the 18th ASITA National Conference*, 14–16 October, Firenze, Italy, pp. 359–363, 2014.

[16] Costabile, S, Martini, S., Petriglia, L., Corrarello, G. & Avanzi, A., Un approfondimento sulle metodologie di conversione e trasformazione coordinate. *Geoportale Nazionale, Geomedia*, **16**(6), pp. 26–28, 2012.
[17] Decreto Legislativo 9 aprile 2008, n. 81. Attuazione dell'articolo 1 della legge 3 agosto 2007, n. 123, in materia di tutela della salute e della sicurezza nei luoghi di lavoro. Gazzetta Ufficiale della Repubblica Italiana, Serie Generale, n. 101 del 30 aprile 2008, Suppl. Ordinario n. 108.
[18] Breiman, L., Friedman, J.H., Olshen, R.A. & Stone, C.J., *Classification and Regression Trees*, Wadsworth International, 1984.
[19] Gariazzo, C., Stafoggia, M., Bruzzone, S., Pelliccioni, A. & Forastiere, F., Association between mobile phone traffic volume and road crash fatalities: A population-based case-crossover study. *Accident Analysis and Prevention*, **115**, pp. 25–33, 2018.
[20] Gariazzo, C. et al., Gli incidenti con mezzo di trasporto. Un'analisi integrata dei determinanti e dei fattori di rischio occupazionali. Collana Ricerche. Istituto Nazionale per l'Assicurazione contro gli Infortuni sul Lavoro (INAIL): Rome, 2019.
[21] Pireddu, A. & Bruzzone, S., Incidenti in gallerie stradali. Istituto Nazionale per l'Assicurazione contro gli Infortuni sul Lavoro (INAIL): Rome, 2019.
[22] Decreto Legislativo 30 aprile 1992, n. 285. Testo aggiornato recante il nuovo codice della strada. Gazzetta Ufficiale della Repubblica Italiana, Serie Generale, n. 67 del 22 marzo 1994, Supplemento Ordinario n. 49.
[23] Lemke, K., Road safety in tunnels. *Transportation Research Record*, **1940**, pp. 170–174, 2000.
[24] Bassan, S., Overview of traffic safety aspects and design in road tunnels. *International Association of Traffic and Safety Sciences (IATSS) Research*, **40**(1), pp. 35–46, 2016.
[25] Nævestad, T.O. & Meyer, S., A survey of vehicle fires in Norwegian road tunnels 2008–2011. *Tunnelling and Underground Space Technology*, **41**, pp. 104–112, 2014.
[26] Xing, Y., Lu, J. & Wang, C., Single vehicle traffic accidents in Shanghai river-crossing tunnels. *Proceedings of the Information Technology and Mechatronics Engineering Conference (ITOEC)*, 2015.

ANALYTICAL MODEL FOR ASSESSING THE RESILIENCE OF CROSS-BORDER CRITICAL TRANSPORTATION INFRASTRUCTURE

FABIO BORGHETTI & GIOVANNA MARCHIONNI
Mobility and Transport Laboratory, Politecnico di Milano, Italy

ABSTRACT

Modern industrialized countries depend on the proper functioning of interdependent infrastructure systems, such as transportation, energy, water, and telecommunications networks. These systems are often critical because they contribute to the organization, functionality, and stability of society. Accidents and disruptions can affect these systems, generating consequences and impacts on the economy, health, safety, and welfare of citizens in one country or several neighboring countries. In case of disruption of a critical cross-border transportation infrastructure, impacts affect not only the area of the event but also a wider area. Depending on the type of event and estimated duration, impacts on the mobility of people and goods can be assessed by considering delays, increased traffic, and potential increase in accidents. The goal of this paper is to present an analytical model to study the resilience of critical road and rail transportation infrastructure; the model is applied to a case study including the Lombardy region (Italy) and Canton Ticino (Switzerland) to verify its validity. The proposed model was developed within the SICt project – Resilience of Cross-Border Critical Infrastructure as part of the Interreg VA Italy–Switzerland Program 2014–2020. The model proposes a resilience index (*RI*) calculated for the road and rail transport network element (link). The formulation of the index considers three independent indicators: (i) *RIRM* (resilience index – rescue management) related to the resources that can be activated and used to cope with an event; (ii) *RIPP* (resilience index – plans and management) related to the rapidity with which the necessary resources can be activated considering the presence of plans and procedures; and (iii) *RIRN* (resilience index – network and traffic) related to the robustness of the transport network elements. This paper focuses on the third indicator (*RIRN*) with reference to the analytical formulation and application of the case study.

Keywords: transport resilience, critical infrastructures resilience, critical infrastructures safety, emergency management, transport vulnerability, road network.

1 INTRODUCTION

Modern, industrialized societies are increasingly dependent on the functioning of interdependent systems and infrastructures, such as information technology, transportation (e.g., road, rail, air and sea), energy and water networks. Dependence is when there is a link or connection between two infrastructures and the state of one infrastructure influences or is related to the state of the other. Interdependence consists of a bidirectional relationship between two infrastructures through which the state of each influences or it is related to the state of the other [1]. These systems and infrastructures are often considered critical because they are necessary for the organization, functioning and stability of a country [2], [3]. Failures, malfunctions, and more generally relevant events of anthropogenic or natural origin make these systems vulnerable, generating consequences for the economy, health, safety, and well-being of the citizens of an entire country or of several neighboring countries [4], [5]. The disruption of a critical transportation infrastructure, road or rail, can have effects not only at the location of the event, but also on a wider area. Depending on the type of event, which can be natural (e.g., landslide, flood, earthquake, etc.) or anthropogenic (e.g., road accident) in origin, and its relative duration, it is possible to estimate the impacts on the mobility of people and goods. Impacts can be analyzed by considering, for instance, delays (alternative

WIT Transactions on The Built Environment, Vol 214, © 2022 WIT Press
www.witpress.com, ISSN 1743-3509 (on-line)
doi:10.2495/SSR220071

routes), increased traffic (congestion), and possible increase in accidents. More generally, socioeconomic impacts on the community can be estimated. At the European level, the Trans-European Network Transport (TEN-T) has been defined as the set of linear (rail, road and river) and punctual (urban nodes, ports, interports and airports) infrastructure considered important at the community level. In practice, nine strategic corridors were identified with the goal of promoting coordinated implementation and development of the Trans-European Transport Network among different states. In addition, European Directive 2008/114/EC on the identification and designation of European critical infrastructures and the assessment of the need to improve their protection was released in 2008; in particular, the energy and transport sectors are considered. To date, an update of the Directive is underway. With this in mind, the aim of this research is to propose and to apply a quantitative method to characterize the links of a road or of a rail transportation network according to a resilience index (RI). This index is composed of three indicators: (i) *RIRM* – rescue management related to the resources that can be activated and used to cope with an event; (ii) *RIPP* – plans and management related to the speed with which the necessary resources can be activated and, in fact, considers management aspects such as the presence of plans and procedures; (iii) *RIRN* – network and traffic related to the robustness of the elements of the transportation network. In this paper, the third *RIRN* indicator is illustrated along with an application case. The paper is organized as follows: Section 2 provides the background with reference to the technical and scientific literature on the issues of vulnerability and resilience; Section 3 illustrates the method from an analytical point of view while Section 4 deals with the application of the method in the Interreg IT-CH SICt project; finally, Section 5 contains the conclusions.

2 BACKGROUND

The topic of vulnerability and resilience of transportation networks is continuously addressed by various authors in technical scientific literature. Definitions of vulnerability and resilience are proposed in the several studies, but to date it is almost impossible to define them in a unique way. However, over the years the scientific community has managed to consolidate some important issues. When discussing critical transportation infrastructure and more generally systems for the mobility of people and goods, it is necessary to define their vulnerability according to potential risks. In this way, decision-making processes that support intervention strategies and policies can be identified and undertaken [6]. More generally, risk can be estimated by considering the probability of occurrence of a specific hazard, identifying the affected elements and assessing their vulnerability to that specific hazard [7], [8]. As a matter of fact, vulnerability is considered an element of risk along with probability and exposure [9]. Making a focus on the transportation sector, vulnerability can be defined as the susceptibility to incidents that may result in significant reductions in road network operations [10], [11]. Vulnerability can also be studied by considering different components: physical, functional, organizational, systemic, and topological [12]. The concept of resilience is addressed by scholars on network systems such as energy, transportation, water and communications [11], [13]. Again, it is difficult to provide an unambiguous definition; however, the term resilience comes from the Latin resilīens-ĕntis, the present participle of resilīre meaning to bounce, to jump back. It may also be interesting to connect it with the Latin meaning of resalio, which describes the act of getting back on the boat overturned by the force of the sea: this interpretation is used in the psychological field to indicate the attitude of going forward without giving up, despite difficulties. Among several definitions acknowledged by the scientific community, resilience can be defined as the ability of a system to withstand a major disruption within acceptable degradation parameters and to recover within acceptable time and cost and compound risks [14], [15]. A system, including one related to transportation and mobility, can be considered resilient if it is characterized by a reduction in (i) the probability of failure/malfunction; (ii) the consequences of failure, considering, for example, loss of life, damage, and economic and social consequences; and (iii) recovery time [16], [17]. Basically, resilience can be defined as the ability of an entity and/or system to anticipate, resist, absorb, respond to, adapt to, and recover from a disturbance and more generally from an event. Fig. 1 shows the operational state (performance) of a transportation system as a function of time, considering a relevant event (red point). The trend of the blue curve represents the evolution of system performance over time following the relevant event and can be a useful tool to support traffic, mobility and emergency management. In addition, the different elements that make up resilience are illustrated.

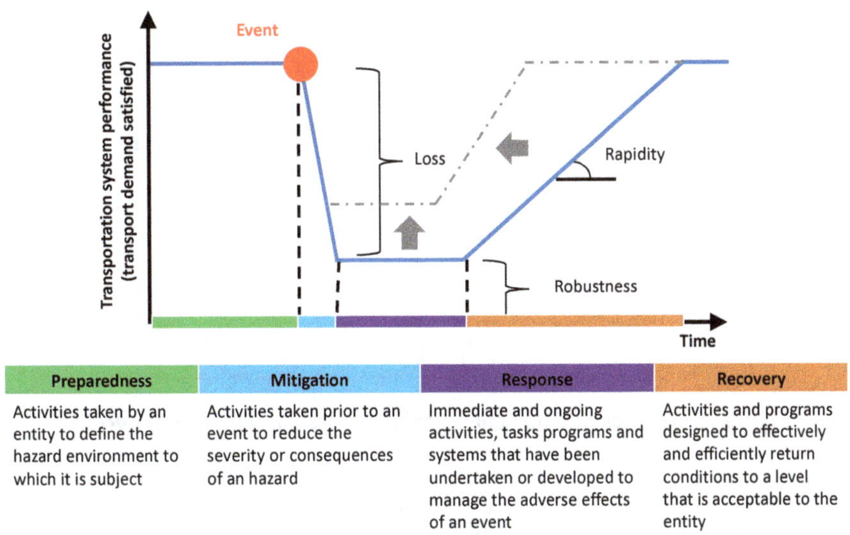

Figure 1: Resilience components for a transport system. *(Source: Adapted from [17], [18].)*

With reference to the management of traffic and mobility of people and goods, it is also important to consider the availability, practicability, and redundancy of routes, which can offer alternatives in the event of a disruption of an element (link). The system can be highly resilient not only if it can resist an event, but also if there are shared procedures and plans for traffic and emergency management. For the management of emergencies following relevant events, it is essential to consider the availability of resources (vehicles, equipment, etc.) that can be activated and used by analyzing, on the one hand, the accessibility of rescue at the location where the event occurred (intervention time) and, on the other hand, redundancy, i.e., how many resources of the same type can be used in the unit of time [12].

3 MODEL DESCRIPTION

A resilience index (*RI*)is calculated for each link *i* of the road and rail transportation network; the Index describes the capacity of each link to cope with a relevant event, whether anthropogenic or natural. *RI* is composed of three independent indicators (see Fig. 2):

$$RI_i = f(RIRM_i; RIPP_i; RIRN_i) \tag{1}$$

where:

- *RIRM* aggregates resilience indices related to the resources that can be managed to cope with a relevant event;
- *RIPP* considers how quickly resources can be activated and managed by assessing the presence, the sharing, and the application of plans and procedures;
- *RIRN* considers the robustness and the importance of network elements based on their relationship to the whole network.

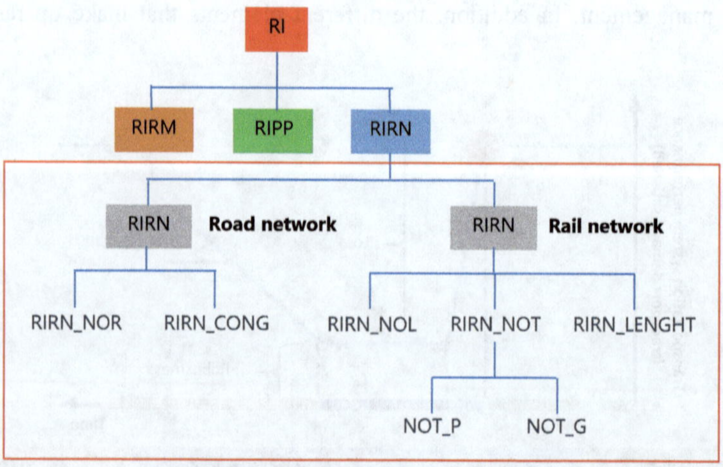

Figure 2: Resilience index analytical structure and focus on the *RIRN*.

The third indicator *RIRN* – network and traffic resilience index, considers the interaction between individual network elements (links) and the network itself both road and rail. In practice, the goal is to assess how important a link is in relation to the entire network if it is not viable following a relevant event. From an analytical point of view, the indicator is calculated differently for the road and rail network. For the road network, the two attributes are considered as follows:

$$RIRN = \mu_1 * RIRN\_NOR + \mu_2 * RIRN\_CONG \quad (2)$$

where:

- *RIRN_NOR* is the number of routes that affect a single link in the road network;
- *RIRN_CONG* represents the traffic/congestion level of the individual road link;
- $\mu 1$ and $\mu 2$ are the relative importance weights of the two indicators. The flexibility of the model allows the analyst to vary the weight and thus the importance of the two contributions.

For the rail network, the three attributes are considered as follows:

$$RIRN = \mu_1 * RIRN\_NOL + \mu_{2rl} * RIRN\_NOT + \mu_{3rl} * RIRN\_LENGTH \quad (3)$$

where:

- *RIRN_NOL* is the number of lines affecting a single link in the rail network;

- *RIRN_NOT* represents the number of trains in the individual rail link;
- *RIRN_LENGTH* represents the relationship between a link and the length of the railway line;
- *μ1, μ2* and *μ3* are the relative importance weights of the two indicators. Again, the flexibility of the model allows the analyst to vary the weight and thus the importance of the two contributions.

3.1 Road network

3.1.1 Calculation of the attribute *RIRN_NOR*

This attribute provides information regarding the number of routes that pass an on single link in the transportation network. Regarding the road network, routes are defined as a sequence of links in the network that join origin–destination (OD) pairs. The calculation of this indicator involves on the one hand zoning the territory (study area) and on the other hand analyzing the routes joining the different OD pairs.

Zoning allows the study of transportation demand: the area is divided into zones and for each zone a centroid is identified that ideally represents the starting and/or ending point of journeys. Zoning is generally the result of a trade-off between quality of analysis and computational burden and still depends on the level of analysis (scale) to be performed; choosing to use smaller zones increases the accuracy of the calculation but also increases the computational burden as there will be more OD pairs to consider. Otherwise, when larger zone sizes are used, the accuracy of the analysis decreases and the computational burden is also reduced. The zoning phase allows the implementation of the OD matrix that in fact represents the transportation demand between each OD pair as illustrated in Fig. 3. Starting from an area (blue line in Fig. 3) it is possible to define zones to which a centroid (blue dot) can be associated. In the example, six zones have been defined, allowing the study of transportation demand (e.g., number of trips, number of passengers, number of vehicles, etc.) using the OD matrix. If the same area had been divided into a larger number of zones, the accuracy would have increased along with the computational burden.

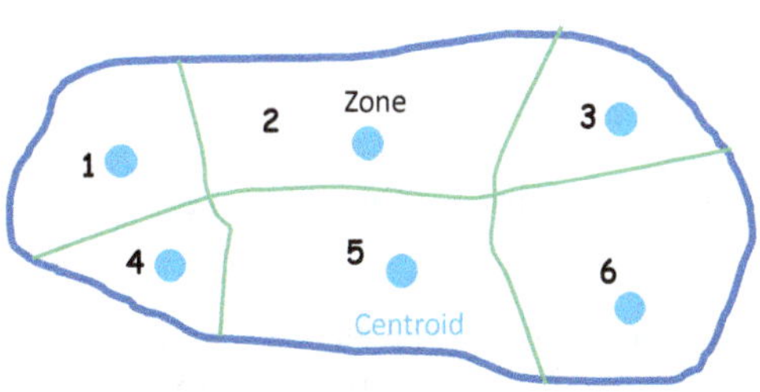

Origin/ Destination	1	2	3	4	5	6
1						
2						
3						
4						
5						
6						

Figure 3: Example of zoning a territory (six zones and six centroids) and OD matrix.

Also, by way of example, it is possible to consider an area divided into four zones as shown in Fig. 4. The blue dots represent the centroids of a zone; the black dots are the nodes and the black arrows are the (oriented) links of the transport network. In fact, the network consists of four nodes (10, 11, 12 and 13) and five links (10–11, 10–12, 11–13, 10–13 and 12–13).

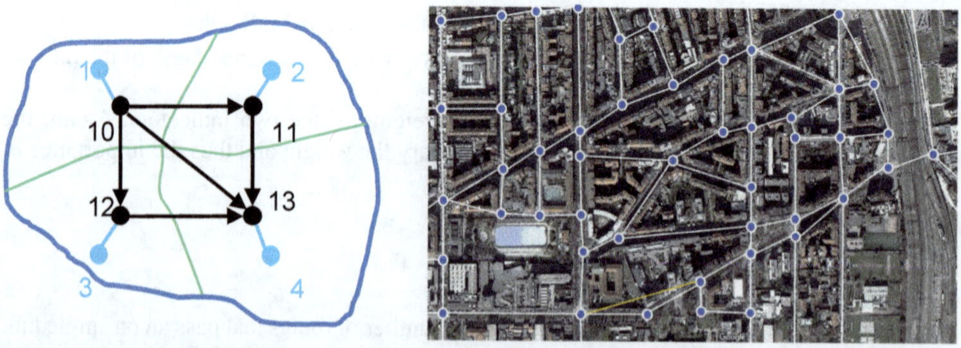

Figure 4: On the left example of four-zone zoning and representation of the transportation network consisting of links and nodes; on the right real example of representation of the transportation network consisting of links and nodes.

With reference to the example in Fig. 4, it is possible to identify seven routes connecting the four zones – OD as shown in Fig. 5. The routes are identified as the paths characterized by the lowest cost to the user considering travel time as the cost.

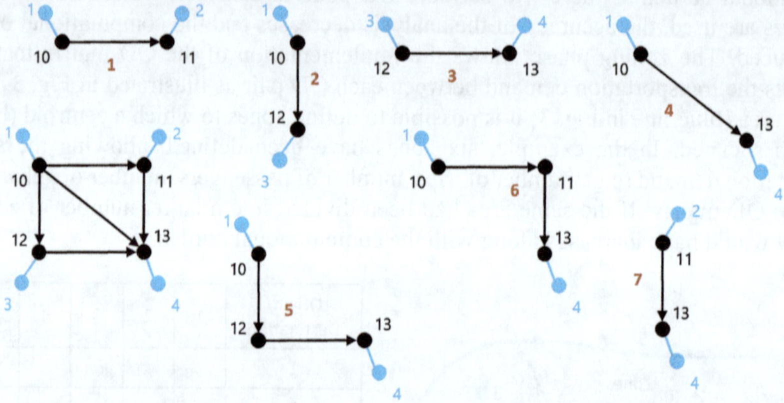

Figure 5: Identification of the seven paths connecting the four OD zones.

In this way it is possible to implement the links–routes incidence matrix where the rows show the links that compose the transportation network (in the example five links) and the columns show the previously identified routes (seven routes) as shown in Fig. 6. For each route the sequence of the links–routes is shown. The values allowed in the matrix are “1” if the link belongs to the generic route or “0” if otherwise.

With this approach, it is possible to know for each link how many paths pass through it; for example, link 12–13 is affected by two paths (3 and 5) while link 10–13 is affected by only one path (4). The next step is to normalize the *RIRN_NOR* attribute using utility functions with a stepwise trend as shown in Fig. 7. On the horizontal axis is the number of paths in each link in the network (*r1, r2, r3, r4*, and *r5*), while on the vertical axis is the normalized value. Five intervals giving rise to five values of the attribute were defined.

Routes

Links

A =

	1	2	3	4	5	6	7
	1-10-11-2	1-10-12-3	1-10-11-2	1-10-13-4	1-10-12-13-4	1-10-11-13-4	2-11-13-4
10-11	1	0	0	0	0	1	0
10-12	0	1	0	0	1	0	0
12-13	0	0	1	0	1	0	0
10-13	0	0	0	1	0	0	0
11-13	0	0	0	0	0	1	1

Figure 6: Example of links–routes incidence matrix.

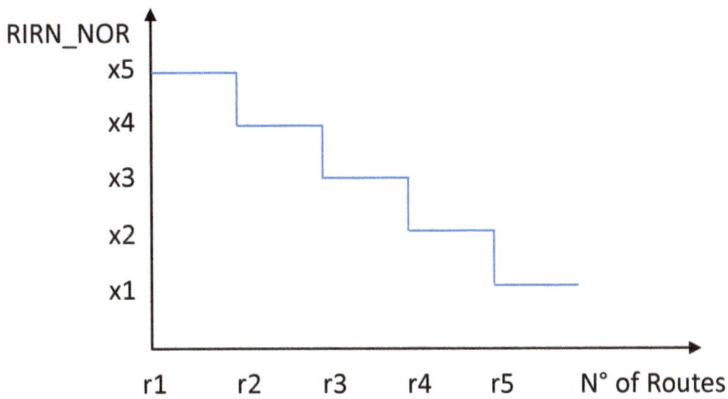

Figure 7: Utility function trend for the calculation of the *RIRN_NOR* attribute.

For instance, the *i*-th link in the transport network may be characterized by a high value of *RIRN_NOR* (x5) because the number of paths passing through it is limited (*r1*); instead, a link characterized by a high number of paths passing through it (*r5*) will be marked by a low value of *RIRN_NOR* (*x1*). In practice, it is reasonable to assume that a link affected by many routes is less resilient because in the event of a disruption, there will be effects and impacts on a larger number of OD pairs; this means that the disruption of the link will have a relevant effect on the network because it will impact on many areas by not guaranteeing connectivity and thus the mobility of people and goods.

3.1.2 Calculation of the attribute *RIRN_CONG*

The second attribute concerns the traffic flow/congestion level of the individual link in the road transport network. If a link has a high level of traffic and/or congestion, it is reasonable to assume that it is an important element in ensuring the mobility of people and goods on the network. In fact, this second attribute can be considered complementary to the previous one concerning the number of routes that affect a link. With reference to the example in Fig. 8 there can be situations where the link (21–24) of the network can be characterized by many routes (four) with low traffic flows (left) or situations where a link is characterized by only one route but with high traffic flows (right). Then there are intermediate situations depending on the configuration of the transportation network.

To consider these two issues simultaneously, the second attribute allows one to assess the presence of high traffic flows or congestion on a single link in the road network. Specifically, the travel time of a link under normal conditions (*Ttrav*) and the travel time during peak hour,

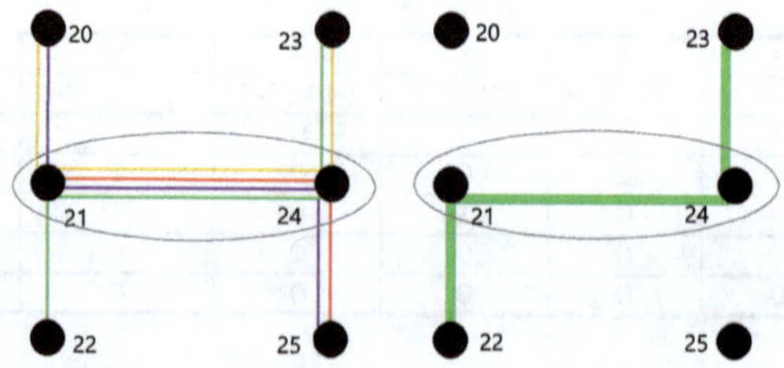

Figure 8: Example of link (21–24) affected by many routes but with limited traffic flow (left) and link affected by only one route characterized by high traffic flow (right).

when the highest traffic flow and possible congestion is assumed (*Tpeak*) are considered. The next step is to determine for each link of the road network the parameter k as follows:

$$k = \frac{T_{peak}}{T_{trav}} \tag{4}$$

If the k value is very high, the link is very congested; on the other hand, if it is equal to 1, it means that during the peak hour the traffic level is not relevant. Again, normalization of the *RIRN_CONG* attribute is performed using utility functions with a stepwise trend as shown in Fig. 7. In practice, it is reasonable to assume that a link affected by a high level of traffic and/or congestion ($k >> 1$) is less resilient since in the event of disruption it will be necessary to manage many vehicles (light and heavy) by taking specific measures such as using viable alternative routes and/or blocking heavy vehicles, etc.

3.2 Rail network

3.2.1 Calculation of the attribute *RIRN_NOL*

This attribute provides information regarding the number of rail lines that pass over a single link in the transportation network. Unlike the road case where the study area has been zoned, it is necessary to consider the geometry of the network and the service provided. For each rail link, it is then possible to determine the number of lines that pass through it. It follows that a link characterized by many lines can be considered less resilient since in the event of disruption it will cause many origins and destinations to be disconnected; this means that the disruption of the link will have a major effect on the network because it will impact many areas by not guaranteeing connectivity and therefore the mobility of people and goods. The next step is to normalize the value of the *RIRN_NOL* attribute using utility functions with a stepwise trend as shown in Fig. 7.

3.2.2 Calculation of the attribute *RIRN_NOT*

The second attribute refers to rail traffic and specifically to the number of trains passing through the individual link in the rail network. Similar to the approach taken for the road network, if a link has a large number of trains, it is reasonable to assume that it is an important element in ensuring the mobility of people and goods on the entire transport network. Again, the attribute can be considered complementary to the previous one concerning the number of lines that affect a link; there may be situations in which a link in the rail network may be characterized by many lines with a reduced number of trains or situations in which a link is characterized by only one line but with a high number of trains. Then there are intermediate situations depending on the configuration of the transportation network. Since in general, the rail network can be used by both passenger and freight trains, the attribute is calculated as follows:

$$NOT \quad = \eta_1 * NOT\_P \quad + \eta_2 * NOT\_G \tag{5}$$

where:

- *NOT* is the number of trains affecting a single link in the rail network;
- *NOT_P* represents the number of passenger service trains affecting a rail link;
- *NOT_G* represents the number of cargo trains affecting a rail link;
- $\eta 1$ and $\eta 2$ are the relative importance weights of the two parameters. Again, the flexibility of the model allows the analyst to vary the weights and thus the importance of the two contributions.

As in the previous cases, it is necessary to normalize the value of the *RIRN_NOT* attribute using utility functions with a stepwise trend as shown in Fig. 6. Again, it is reasonable to assume that a link affected by many trains is less resilient because in the event of a disruption it will be necessary to handle many trains (passenger and freight) by adopting specific measures planned by the service operator (e.g., bus bridging for passenger service).

3.2.3 Calculation of the attribute *RIRN_LENGTH*

The third attribute concerns the relationship between the railroad link on which the relevant event is assumed and the length of the line. The longer a line is, the more reasonable it is to assume that the disruption of the link belonging to the line will cause impacts and consequences on the entire rail network. In fact, longer lines are desired to be penalized, which, in the event of link disruption due to an event, may suffer delays and consequences even over long distances. Again, it is necessary to normalize the value of the *RIRN_LENGTH* attribute using utility functions with a stepwise trend as shown in Fig. 7.

4 CASE STUDY: THE INTERREG SICT PROJECT

The SICt – Resilience of Cross-Border Critical Infrastructure project is developed under the Interreg V-A Italy–Switzerland 2014–2020 cooperation program that contributes to the goals of the Europe 2020 strategy. The project area includes the Lombardy Region in Italy and the Ticino Canton in Switzerland, as illustrated in Fig. 9 and is organized into six work packages (WP). The main goal of the project is to increase knowledge and information sharing on critical cross-border infrastructure with reference to transport infrastructure. These infrastructures represent important and strategic corridors for the mobility of people and goods also at the European level. Three specific goals have been identified for the implementation of the project: (i) to increase and improve cross-border cooperation in terms of governance to manage anthropogenic and natural events that may affect critical infrastructure (road and rail); (ii) to strengthen joint capacities for managing the impacts caused by disruption of cross-border critical infrastructure that may cause effects on both countries (emergency and traffic management); and (iii) to verify and to improve the effectiveness of the cooperation system for monitoring and managing relevant events in the cross-border area (macro-regional area).

Events affecting road and rail transportation infrastructure are considered in the study area; impacts of events occurring within the study area are evaluated in the impact area (see Fig. 10). Some representative results of *RIRN* related to the road and rail network are shown in Fig. 11. Links in green color have a higher indicator value than those in black color; therefore, they are found to be more resilient.

Figure 9: Main information and area of the Interreg SICt project.

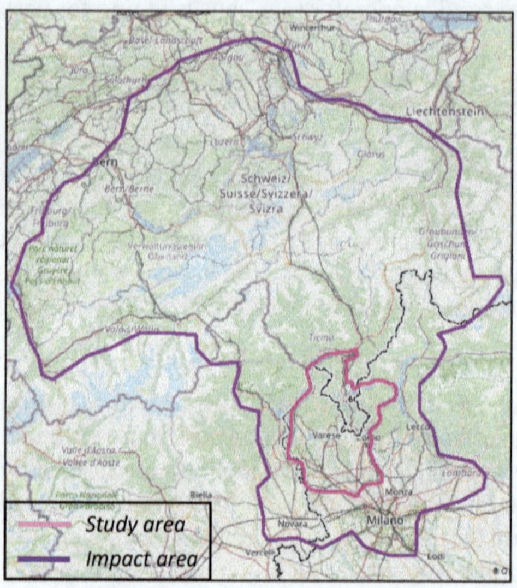

Figure 10: Study area and impact area of the Interreg SICt Project.

5 CONCLUSION

The paper presents the analytical formulation and application of a Resilience Index (*RI*), determined for each element of the road and rail network. The proposed approach can be considered as a useful decision support system (DSS) that can be used by those involved in the decision-making process of planning and managing the resilience of transportation infrastructure (infrastructure managers, first responders, administrations, etc.).

The method can be used in two phases: (i) planning phase, when it is necessary to simulate the joint response of the usable and available resources and each operator can share and know the critical points of the transportation network; (ii) emergency phase, to provide stakeholders

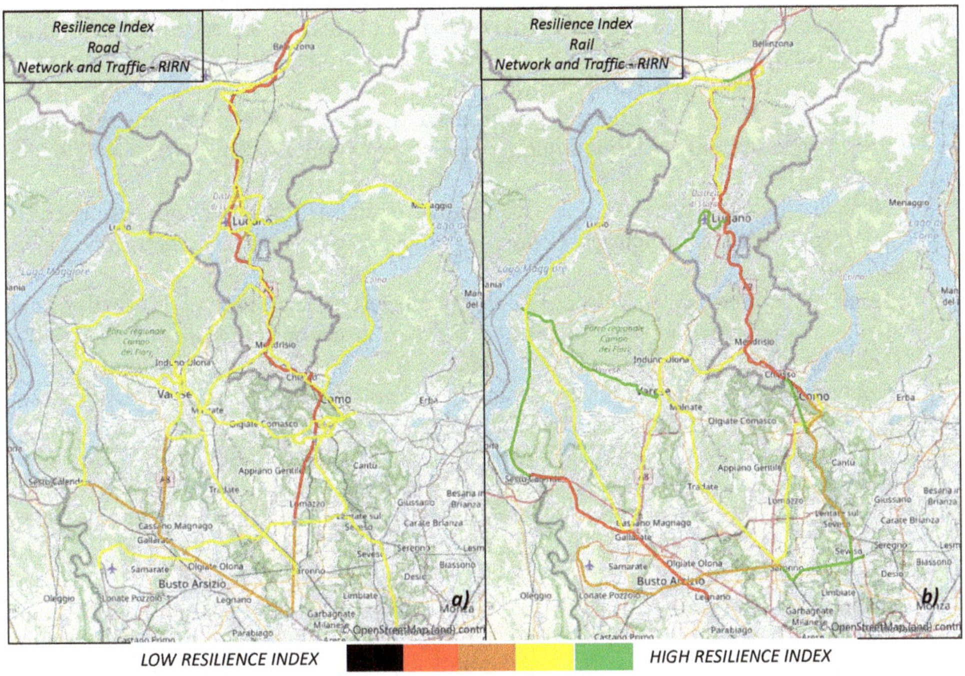

Figure 11: (a) *RIRN* of the road network; and (b) *RIRN* of the rail network.

with an assessment of the impact of the event on the network based on the available resources and the condition of the network itself. By comparing the different links that make up the transportation network, it is possible to prioritize interventions, identifying situations where it is most urgent to allocate resources to improve resilience by acting on different indicators and attributes. The strengths are many: the proposed method is replicable, modular and expandable; it can be replicated in other contexts at different levels of detail depending on the analyst's needs: for example, at local, provincial, regional, national and macro-regional scales. It also allows for the graphical representation of specific indicators and attributes and enables the consideration and inclusion of new parameters for resilience assessment. From an analytical point of view, the calculation process does not require special resources, making the method streamlined, operational and easily usable. Possible developments of the work include the dependence of the method on the data needed; hence the importance of the quality of data and information such as (i) resource mapping, (ii) transportation graph, (iii) traffic data, (vi) availability of emergency and traffic management plans and procedures, etc.

REFERENCES

[1] Rinaldi, S.M., Peerenboom, J.P. & Kelly, T.K., Identifying, understanding, and analyzing critical infrastructure interdependencies. *IEEE Control Systems Magazine*, **21**(6), pp. 11–25, 2001. DOI: 10.1109/37.969131.

[2] Cohen, F., What makes critical infrastructures critical? *International Journal of Critical Infrastructure Protection*, 53–54, 2010.

[3] European Commission, Council Directive 2008/114/EC of 8 December 2008 on the identification and designation of European critical infrastructures and the assessment of the need to improve their protection, 2008.

[4] Eusgeld, I., Nan, C. & Dietz, S., "System-of-systems" approach for interdependent critical infrastructures. *Reliability Engineering and System Safety*, **96**(6), pp. 679–686, 2011.

[5] Nan, C. & Sansavini, G., A quantitative method for assessing resilience of interdependent infrastructures. *Reliability Engineering and System Safety*, pp. 35–53, 2017.

[6] Hong, W., Clifton, G. & Nelson, J.D., Rail transport system vulnerability analysis and policy implementation: Past progress and future directions. *Transport Policy*, 2022. DOI: 10.1016/j.tranpol.2022.02.004.

[7] Coburn, A.W., Spence, R.J.S. & Pomonis, A., *Vulnerability and Risk Assessment*, 2nd ed., UNDP Disaster Management Training Programme, 1994.

[8] Linkov, I., Fox-Lent, C., Read, L., Allen, C.R., Arnott, J.C., Bellini, E., Woods, D., Tiered approach to resilience assessment. *Risk Analysis*, **38**(9), pp. 1772–1780, 2018.

[9] UNDRO – Office of the United Nations Disaster Relief Coordinator, Natural disasters and vulnerability analysis. Report of Expert Group Meeting, 9–12 Jul., Geneva, Switzerland, 1979.

[10] Berdica, K., An introduction to road vulnerability: What has been done, is done and should be done. *Transport Policy*, pp. 117–127, 2002.

[11] Mattsson, L. & Jenelius, E., Vulnerability and resilience of transport systems: A discussion of recent research. *Transportation Research Part A: Policy and Practice*, pp. 16–34, 2015.

[12] Borghetti, F., Petrenj, B., Trucco, P., Calabrese, V., Ponti, M. & Marchionni, G., Multi-level approach to assessing the resilience of road network infrastructure. *International Journal of Critical Infrastructures,* **17**(2), pp. 97–132, 2021.

[13] Reggiani, A., Network resilience for transport security: Some methodological considerations. *Transp. Policy*, **28**, pp. 63–68, 2013.

[14] Francis, R. & Bekera, B., A metric and frameworks for resilience analysis of engineered and infrastructure systems. *Reliability Engineering and System Safety*, pp. 90–103, 2014.

[15] Haimes, Y.Y., On the definition of resilience in systems. *Risk Analysis*, pp. 498–501, 2009.

[16] Bruneau, M., Chang, S.E., Eguchi, R.T., Lee, G.C., O'Rourke, T.D., Reinhorn, A.M. & Von Winterfeldt, D., A framework to quantitatively assess and enhance the seismic resilience of communities. *Earthquake Spectra*, pp. 733–752, 2003.

[17] Carlson, L., Haffenden, J.A., Bassett, R., Buehring, G., Collins, W.J., Folga, M.S., Petit, F., Phillips, J.A., Verner, D.R. & Whitfield, R.G., *Resilience: Theory and Application*, 2012.

[18] Deublein, M., Roth, F., Willi, C., Anastassiadou, K. & Bergerhausen, U., Linking science to practice: A pragmatic approach for the assessment of measures to improve the resilience of transportation infrastructure systems. *Proceedings of the 29th European Safety and Reliability Conference*, pp. 1351–1356, 2019.

SECTION 3
CYBERSECURITY/ E-SECURITY

BUILDING VULNERABILITY ASSESSMENT FOR EXPLOSIVE AND CBR TERRORIST ATTACKS

MARCO CARBONELLI[1], RICCARDO QUARANTA[1], ANDREA MALIZIA[2], PASQUALE GAUDIO[1], DANIELE DI GIOVANNI[1,3] & GRACE P. XERRI[1]
[1]Industrial Engineering Department, University of Rome Tor Vergata, Italy
[2]Department of Biomedicine and Prevention, University of Rome Tor Vergata, Italy
[3]Unicamillus-Saint Camillus International University of Health Sciences, Italy

ABSTRACT

Assessing the vulnerabilities of a building/site for a specific threat is one of the key issues in the risk assessment process. A vulnerability is defined as any weakness that can be exploited by an aggressor to make an asset susceptible to damage. The purpose of the vulnerability assessment process discussed in this paper is to identify the main vulnerabilities which influence a building's risk level when a specific explosive or chemical, biological, radiological (CBR) threat arises. Vulnerability assessments are designed to provide an in-depth analysis of the characteristics of a facility and its associated elements to identify building weaknesses and lack of redundancy, as well as to determine protective or corrective actions that can be designed or implemented to reduce building vulnerabilities. This work proposes an innovative building vulnerability assessment method (BVAM), comprised of three steps. The first step, building criticality analysis (BCA), seeks to verify the criticality of several building aspects elaborated from best practices on the analysis of building structure and function. The result of this BCA determines if critical building components or systems, designed for the deterrence, detection, and limitation of damages, can continue to function properly during a crisis, and to ensure the correct operation of the emergency systems. The second step aims at characterising the application of a given number of specific threats to the building. The third step focuses on a final assessment of the level of vulnerability associated with the various applied threats, for the specific building and the specific assets to be protected. This result is achieved by employing a proposed seven-level vulnerability scale. The result of the evaluation of the level of vulnerability can be used for the final risk assessment phase.
Keywords: risk assessment, vulnerability assessment, buildings, terrorism, explosive, unconventional attacks, CBR.

1 INTRODUCTION

Many definitions of risk are available in the technical literature [1]–[6]. In any of these works, the concept of risk is always associated with uncertainties related to future events.

In practice, risk is a hazard or an exposure to a possibility of loss or damage, or ability to suffer a possible loss [4]. The estimation of risk [2] is usually found by the probability of the event occurring multiplied by the consequence of the event, given that it has occurred. In other words, risk is considered as a combination of the consequences of an event and the associated likelihood/probability of its occurrence [1], [7], [8]. Hereafter, three approaches to the risk assessment are briefly described.

According to the USA DHS [9], risk "R" is mathematically expressed as a function of the threat probability "T" to a target/area, the vulnerability "V" of the target/area, and the consequence "C" of an attack on that target/area, as described in eqn (1):

$$R = f(T, V, C). \quad (1)$$

In the approach proposed by the UN [10], risk "R" is expressed as a function of hazard probability "H", vulnerability "V" and exposure "E", as described in eqn (2):

$$R = f(H, V, E). \quad (2)$$

WIT Transactions on The Built Environment, Vol 214, © 2022 WIT Press
www.witpress.com, ISSN 1743-3509 (on-line)
doi:10.2495/SSR220081

Finally, in the European approach [9], risk "R" is a function of the probability of occurrence of a hazard "P", the exposure "E" (total value of all elements at risk), and the vulnerability "V", as described in eqn (3):

$$R = f(P, V, E). \tag{3}$$

The EU technicians highlight that the impacts of a hazard are also a function of the preventive and preparatory measures that are employed to reduce the risk. In other words, effective prevention and preparedness measures can decrease the vulnerability and therefore the risk.

As a general statement, vulnerabilities are the characteristics of an asset, system, location, process, or operation that render it susceptible to destruction, incapacitation, or exploitation by mechanical failures, natural hazards, terrorist attacks, or other malicious acts. A vulnerability can therefore be defined as any weakness that can be exploited by an aggressor to make an asset susceptible to damage.

Based on this consideration and the approaches abovementioned, assessing the vulnerabilities of a building for a specific threat is one of the key issues in the risk assessment process.

Vulnerability assessments (VAs) for buildings [11] are designed to provide an in-depth analysis of the characteristics of the facility or its associated elements to identify weaknesses and lack of redundancy, as well as to determine protective or corrective actions that can be designed or implemented to reduce the vulnerabilities.

2 OBJECTIVES

In this work, an original building vulnerability assessment method (BVAM) is proposed. The purpose of this VA process is to identify the vulnerabilities that mainly influence the level of risk of a building when a specific explosive or CBR threat arises. The method proposed is based on an analytical procedure structured around 76 different items organised into nine topics, which include physical and organisational aspects and social, economic, structural and institutional factors, with the aim of identifying the building criticalities. The result of the BVAM is based on a seven-level vulnerability scale and will provide numerical values that represent, for the scenario analysed, different levels of vulnerability. The numerical value of the vulnerability level thus assessed can be used, in combination with the values of threat level and exposure level, in the calculation of the level of risk associated with a building.

3 BUILDING VULNERABILITY ASSESSMENT METHOD

The BVAM proposed in this paper, has been developed and adapted from the USA Department of Veterans Affairs checklist [11] and from the risk analysis model presented in Carbonelli [1].

The method is structured in three different steps, as represented in Fig. 1.

- Step 1: Proposes to verify, through the building criticality analysis (BCA), the criticality of 76 items, grouped into nine topics, elaborated from the best practices on the analysis of building structure and functions.
- Step 2: Aims at characterising a given number of specific threats to be applied to the building.
- Step 3: Focuses on the final assessment of vulnerability associated with the specific considered threats, for the specific building, and for the specific assets to be protected, using a proposed Vulnerability Scale comprised of 7 levels.

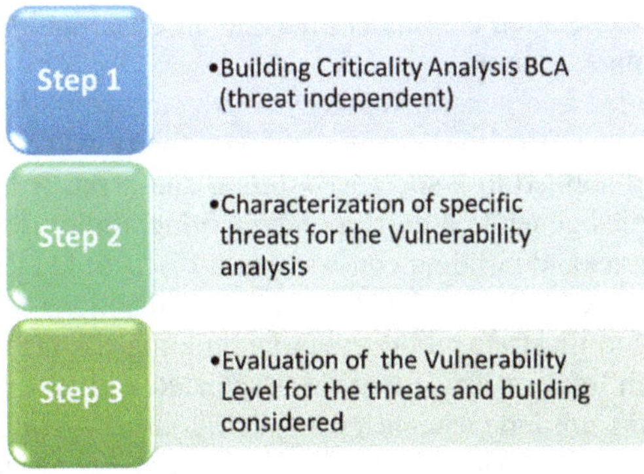

Figure 1: Building vulnerability assessment method in three steps.

The result of the VA provides a numerical value which can be used in the final risk assessment.

3.1 Step 1: Building criticality analysis (BCA)

The BCA proposed in Step 1 can be considered as a preliminary assessment of the weaknesses of different aspects of the building site, structure, and functions. In addition, this analysis allows for the evaluation of design issues that could potentially reveal exploitable vulnerabilities.

The result of the analysis determines if, during a crisis, critical components/systems will continue to work properly in order to enhance deterrence, detection, and limitation of damages, and to ensure the correct operation of the emergency systems. The BCA is categorised in nine sections, indicated as "topics", listed in Table 1.

Table 1: Criticality topics and number of items per topic (Step 1).

Topic	Criticality topics	No. of items
1	Site characteristics	12
2	Architecture	10
3	Structural systems	7
4	Building envelope	5
5	Utility systems	8
6	Mechanical systems and HVAC	10
7	Infrastructure and systems of internal essential services (plumbing, gas systems, electrical power, fire alarms, telephone and ICT services)	11
8	Security systems	8
9	Emergency, security and operation continuity plans	5
Total	Nine topics	76 items

To conduct a complete building VA, each topic should be assigned to the identified assessment team (AT). Such a team should be composed by engineers, architects, or subject matter experts who are knowledgeable and qualified to perform an accurate analysis. The AT should carefully analyse the topics in Table 1 (from the site characterisation to the emergency, security and operation continuity plans) in order to highlight possible criticalities and potential related vulnerabilities.

A criticality is intended as a general weakness that could be potentially exploited for an attack. In this approach, a criticality becomes a vulnerability when a detailed and specific threat is considered and applied to a specific building and asset. It is important to observe that not all the criticalities generate a correspondent vulnerability; this correlation depends on the specific threat, asset and building considered, as discussed below.

The nine topics suggested reflect different aspects and functions typical of a building; the objective of the BCA is to illustrate all the essential building characteristics to determine an accurate result. For each topic, a list of items – associated with one or more questions – is included. These 76 items are independent of a specific threat and must be considered and evaluated by the AT through a criticality scale (Table 2) to determine their criticality. The criticality evaluation of a single item is carried out by adopting a four-level scale based on a quantitative weight score (WS). For this scale, the tripling criteria is applied. The rationales for adopting a quantitative scale based on the tripling criteria is widely discussed in Carbonelli [1].

Table 2: Criticality scale for item analysis on four levels.

Criticality scale (for items)	Criticality WS
Extreme	27
Elevated	9
Marginal	3
Negligible	1
Not applicable	–

As an example, one of the 76 different criticality scales used in this method is reported in Table 3, that responds to the question: *Are there any major/critical infrastructures surrounding the building*?

Table 3: Criticality scale relative to item 1.1 (Surrounding structures/facilities).

1.1: Criticality scale (Surrounding structures/facilities)		WS
Extreme	Many significant critical infrastructures are adjacent to the main building considered	27
Elevated	Some significant critical infrastructures are adjacent to the main building considered	9
Marginal	No major critical infrastructure and only infrastructures of secondary importance are adjacent to the main building considered	3
Negligible	No significant infrastructure is adjacent to the main building considered	1
N/A	Not applicable: it is not possible to give a relevant answer to this question	–

Using these scales, the AT can provide for each item a relevant WS to highlight the criticality conditions for the final VA.

For each topic, a cumulative criticality evaluation is then obtained by calculating for all the WSs, the average "m", the standard deviation "s" [12], and the modified average "m_{mod}", as defined in eqn (4):

$$m_{mod} = m + s. \quad (4)$$

Below, one of the nine topic tables used in the BCA is presented as an example (Table 4). Each line of every table represents a specific item and the questions that the AT must answer to evaluate the criticality, taking into account the relative criticality scale. Each of the nine tables enables the evaluation of the relative items expressed by a criticality WS assigned by the AT, and the average and standard deviation of the WSs. These values provide a rapid indication of the general criticality of the topic.

Table 4: Topic 1 – Site characteristics, table of items.

Topic 2: Architecture			
Item no.	**Item**	**Question**	**WS**
1.1	Surrounding structures/facilities	Are there major/critical infrastructures surrounding the building?	
1.2	Terrain characteristics	Does the terrain place the building in a depression or low area?	
1.3	Curb lane parking characteristics	Is curb lane parking for unmonitored parked vehicles unacceptably close to the building?	
1.4	Perimeter barriers for pedestrian access	Is a perimeter fence, or other types of barrier controls, in place for the pedestrian access?	
1.5	Vehicle access points	Are the vehicle access points well designed?	
1.6	Pedestrian access control	Is pedestrian access controlled at the perimeter of the building?	
1.7	Private vehicle access control	Is private vehicle access controlled at the perimeter of the building?	
1.8	Shipping/delivery vehicle access control	Are shipping and delivery vehicles controlled at the building entrance?	
1.9	Alternative potential access	Are there any exploitable potential access points to the building through utility paths or water runoff?	
1.10	Anti-ram devices	What are the types of vehicle anti-ram devices at the building?	
1.11	Site lighting in the external area	Is the site lighting adequate from a security perspective in roadway access and parking areas?	
1.12	External connection to the building	Are any of the nearby in-ground and out-ground infrastructures directly connected to the building?	
Topic 1			
Average of criticality WSs			
Standard deviation of criticality WSs			

3.2 Step 2: Characterisation of specific threats for the VA

Once the general criticalities of the building have been examined, it is necessary to introduce and describe the specific threats deemed to be more likely applied to the building under assessment. In Step 2 of this BVAM method, a specific VA is carried out by the AT that performs the following activities, considering the selected threats:

- for each selected threat, the agent/explosive and vector types, the possible maximum size/quantity of the agent/material used in the attack, and the possible specific locations in the building are analysed in detail.
- the BCA results obtained in Step 1 provide immediate indications of the weaknesses that can be exploited, becoming effective vulnerabilities. These indications provide crucial elements to mitigate the vulnerability, by reducing the associated criticality, and assessing the specific vulnerability of the building (Step 3) related to the considered threats.

3.3 Step 3: Evaluation of the vulnerability level (VL) for the building

At the end of Step 2, the AT has a clear picture of the exploitable criticalities of the building with respect to the threats and the assets considered. The overall VL is assessed in this Step 3. For each threat considered, the AT evaluates a specific building VL, using a seven-level vulnerability scale which provides qualitative and quantitative definitions for each level. The vulnerability scale is described in Table 5. The seven levels proposed in the table represent seven contiguous ranges of vulnerability in the interval from 0 to 1, where 0 represents the minimum vulnerability value (i.e., totally invulnerable) and 1 represents the maximum vulnerability value (i.e., totally vulnerable).

Following the method proposed, the AT has, at this point, a clear picture of the building criticalities and the threats to be applied. Only under this condition is it possible to provide a reliable evaluation of the specific VL. This proposed vulnerability scale provides not only qualitative descriptions of the VL but also measurable quantitative values and adopts a logarithmic approach for the definition of the range of each level. This type of approach, as discussed in Carbonelli [1], has many advantages over a linear approach. The quantitative value can be used in the calculation of the overall risk associated with the building.

4 BVAM APPLICATION TO A CASE STUDY

In order to render the BVAM more tangible, a practical application of the proposed method has been carried out by analysing a real shopping centre to establish a relevant case study. The shopping centre, whose exact information is not disclosed for security purposes, is located in the outskirts of an important Italian town. The following three scenarios have been considered:

- the explosion of a suicide belt bomb.
- the explosion of a van bomb.
- the explosion of a Caesium-137 dirty bomb.

4.1 Building criticality analysis (BVAM Step 1)

The data of this study has been collected through an inspection of the shopping centre, with permission from the property. The results were processed using a prototype BCA software tool developed on a spreadsheet application specifically for this work. The results obtained for this case study are reported in ten tables, of which one is presented as an example (Table 6). The criticality WS values were entered by the AT. The quantitative WS values correspond, instead, to the automatic data processing of the software tool.

Table 5: Seven level vulnerability scale.

Vulnerability rating	Qualitative	Quantitative (no. of successes/no. of attempts)	Level description
7	Very high	From 3^{-1} to 3^{0} (1/3–1)	One or more major vulnerabilities that make the asset extremely susceptible to an aggressor, for the specific threat considered. The building lacks redundancies/physical protection/ resilience. The entire building would only be functional again a very long period of time after an event.
6	High	from 3^{-2} to 3^{-1} (1/9–1/3)	One or more major vulnerabilities that make the asset highly susceptible to an aggressor, for the specific threat considered. The building has poor redundancies/physical protection/ resilience, and most parts of the building would only be functional again a long period of time after an event.
5	Medium high	from 3^{-3} to 3^{-2} (1/27–1/9)	An important vulnerability that makes the asset very susceptible to an aggressor, for the specific threat considered. The building has inadequate redundancies/physical protection/resilience, and most critical functions would only be operational again a long period of time after an event.
4	Medium	from 3^{-4} to 3^{-3} (1/81–1/27)	A vulnerability that makes the asset fairly susceptible to an aggressor, for the specific threat considered. The building has insufficient redundancies/physical protection/ resilience, and most parts of the building would only be functional again a considerable period of time after an event.
3	Medium low	from 3^{-5} to 3^{-4} (1/243–1/81)	A vulnerability that makes the asset somewhat susceptible to an aggressor, for the specific threat considered. The building has a fair level of redundancies/physical protection/ resilience, and most critical functions would only be operational again a considerable period of time after an event.
2	Low	from 3^{-6} to 3^{-5} (1/729–1/243)	A minor vulnerability that slightly increases the susceptibility of the asset to an aggressor, for the specific threat considered. The building has a good level of redundancies/physical protection/resilience, and the building would be operational within a short period of time after an event.
1	Very low	$< 3^{-6}$ (< 1/729)	No relevant vulnerability appears after the analysis. The building has excellent redundancies/physical protection/resilience, and the building would be operational immediately after an event.

Table 6: Case study results of the BCA for Topic 3 – Structural systems.

Topic 3: Structural systems				
Item no.	**Item**	**Questions**	**Criticality WS**	**Quantitative WS**
3.1	Construction characteristics	What type of construction? What type of concrete and reinforcing steel? What type of steel? What type of foundation?	Marginal	3
3.2	Structural and non-structural components	Are any of the structural/non-structural components vulnerable either directly or indirectly to explosive blast?	Elevated	9
3.3	Progressive collapse	Is the building capable of sustaining the removal of a column for one floor above grade at the building perimeter without progressive collapse?	Marginal	3
3.4	Floor of loading dock	Will the loading dock design limit damage to adjacent areas and vent explosive force to the exterior of the building?	Extreme	27
3.5	Mailroom explosion mitigation	Are mailrooms, where packages are received and opened for inspection, and unscreened retail spaces designed to mitigate the effects of a blast on primary vertical or lateral bracing members?	Elevated	9
3.6	In-ground structural systems	Would failure of part of the in-ground infrastructure affect the structural system of the building?	Elevated	9
3.7	Underground water presence	Does the presence of underground water under the building generate instability and unacceptable flooding?	Elevated	9
Topic 3				
Average of criticality WSs			9.86	
Standard deviation of criticality WSs			7.47	

The total results of the BCA of the shopping centre, for all the nine topics, are summarised in Table 7.

Table 7: Summary of the results obtained for the shopping centre case study.

Topic criticality analysis				
Topic no.	**Topic name**	**m**	**s**	**m_{mod}**
1	Site characteristics	7.00	6.93	13.93
2	Architecture	13.67	9.71	23.37
3	Structural systems	9.86	7.47	17.33
4	Building envelope	11.40	8.14	19.54
5	Utility systems and internal distribution infrastructures	6.75	7.9	14.65
6	Mechanical systems – HVAC	15.00	10.04	25.04
7	Infrastructures and systems of internal essential services	4.09	2.31	6.41
8	Security systems	10.33	6.18	16.52
9	Emergency, security and operation continuity plans	7.80	2.40	10.20

The m_{mod} index can be interpreted using a final criticality scale (Table 8).

Table 8: Criticality Scale based on m_{mod}.

Criticality m_{mod} Scale	**Range**
Extreme	>15
Elevated	7–15
Marginal	3–6.99
Negligible	1–2.99
NA	–

Based on Table 8, the analysis of the results from Table 7 highlights that:

- Topics #2, 3, 4, 6, 8 show an extreme criticality.
- Topics #1, 5, 7, 9 show an elevated criticality.
- Topic #7 shows a marginal criticality.

These results indicate a high level of criticality of the building due to the weaknesses identified through the 76 items analysed.

4.2 Characterisation of specific threats (BVAM Step 2)

As abovementioned, three threats were considered in this phase. For each threat, the AT must specify in detail the following:

- the type of agent/explosive.
- the type of vector for the agent/explosive.
- the possible maximum size/quantity of the agent/material.
- the possible specific location, with respect to the building, where the threat might be applied.

Specific and detailed information on different types of explosion and blast characteristics [13] can be found also in a recent European Commission JRC technical report [14] and in the USA FEMA “Reference manual to mitigate attacks against buildings” [11].

Tables 9–11 summarise the assumptions made by the AT in this phase.

Table 9: Characterisation of the threats for the "suicide belt bomb" case.

Case: Suicide belt-bomb	Specific data
Type of agent/explosive	TNT
Type of vector	Belt-bomb
Maximum size/quantity of the agent/material	5 kg
Specific location, with respect to the building, where the threat might be applied	Immediately inside the building from shopping centre entrance

Table 10: Characterisation of the threats for the "van bomb" case.

Case: Van bomb	Specific data
Type of agent/explosive	TNT
Type of vector	Van
Maximum size/quantity of the agent/material	800 kg
Specific location, with respect to the building, where the threat might be applied	Area of access for shipping/delivery vehicles

Table 11: Characterisation of the threats for the "Caesium-137 dirty bomb" case.

Case: Caesium-137 dirty bomb	Specific data
Type of agent/explosive	TNT and Caesium-137
Type of vector	Pick-up truck
Maximum size/quantity of the agent/material	400 kg TNT and 90 g Caesium-137
Specific location, with respect to the building, where the threat might be applied	In the external parking area

Finally, a further evaluation of the criticality items of Step 1 was carried out with the aim of highlighting both the primary weaknesses that can be directly exploited as actual vulnerabilities for threat under analysis, and the secondary weaknesses that, in an indirect manner, contribute to making the consequences of the attack more severe. These were noted with a criticality level as "elevated" or "extreme" in Step 1.

If mitigation actions against the vulnerabilities are to be taken by the AT, the primary vulnerabilities should be reduced first and, only if adequate resources are available, the secondary vulnerabilities should be addressed.

4.3 Evaluation of vulnerability level (BVAM Step 3)

Considering the results obtained in the two previous steps, it was possible for the AT to evaluate the specific vulnerability of the building.

The three threats considered present elevated or extreme VL. In the "suicide belt bomb" case, the area of concern relating to impacts on human health involved the building's internal area. For the other two scenarios, the greatest impacts were found in the external areas of the shopping centre, with maximum consequences in terms of area impacted in the "dirty bomb" explosion case.

Using Table 5, the AT was able to determine the vulnerability rating for the three threats, for example, by assigning a VL of 7 (very high) to the three considered cases.

Table 12: Main exploitable vulnerabilities in case of a "suicide belt bomb" explosion.

Criticality item – Exploitable vulnerability (suicide belt-bomb)	Level of criticality	Vulnerability type
1.6: No pedestrian access control at the perimeter of the building	Elevated	Primary
2.4: Public and employee entrances do not include equipment for access control-screening	Extreme	Primary
2.7: Critical assets (people, activities, building systems and components) are not well separated from main entrance, vehicle circulation, parking	Elevated	Secondary
2.10: Ceiling, internal walls, overhead utilities and lighting systems are not designed to remain in place without generating debris in hazardous events	Extreme	Secondary
4.3: Glazing of the building are not secure in case of blast	Elevated	Secondary
5.6: No redundant and reliable electrical service source	Extreme	Secondary
7.11: No mass notification system that reaches all building occupants	Elevated	Secondary
8.1: CCTV cameras used, 24 hours/7 days a week recorded and monitored at the perimeter and in the critical areas of the building are insufficient	Elevated	Secondary
8.2: Video quality not adequate both in daylight and darkness	Elevated	Secondary
8.6: Security scanners (x-ray, magnetometer, magnetic imaging, etc.) are not used for security purposes in some areas of the building	Extreme	Primary
9.4: Emergency plan not up-to-date and not well designed	Elevated	Secondary
9.5: Operational continuity plan to apply no up-to-date and well-designed	Elevated	Secondary

Table 13: Main exploitable vulnerabilities in case of a "van bomb" explosion.

Criticality item – Exploitable vulnerability (van-bomb)	**Level of criticality**	**Vulnerability type**
1.8: No vehicle access control at the shipping/delivery entry	Elevated	Primary
2.5: Doors and walls along the security screening not adequately reinforced	Elevated	Secondary
2.7: Critical assets (people, activities, building systems and components) are not well separated from main entrance, vehicle circulation, parking	Elevated	Secondary
2.10: Ceiling, internal walls, overhead utilities and lighting systems are not designed to remain in place without generating debris in hazardous events	Extreme	Secondary
4.1: Low designed or estimated protection level of the building envelope against a possible high magnitude explosive threat	Extreme	Secondary
4.3: Glazing of the building are not secure in case of blast	Elevated	Secondary
4.4: Building is not designed to resist to high external pressure (ex. blast)	Elevated	Secondary
5.6: No redundant and reliable electrical service source	Extreme	Secondary
7.11: No mass notification system that reaches all building occupants	Elevated	Secondary
8.1: CCTV cameras used, 24 hours/7 days a week recorded and monitored at the perimeter and in the critical areas of the building are insufficient	Elevated	Secondary
8.2: Video quality not adequate both in daylight and darkness	Elevated	Secondary
9.4: Emergency plan not up-to-date and not well designed	Elevated	Secondary
9.5: Operational continuity plan to apply no up-to-date and well-designed	Elevated	Secondary

Table 14: Main exploitable vulnerabilities in case of a "Caesium-137 dirty bomb" explosion.

Criticality item – Exploitable vulnerability (Caesium-137 dirty bomb)	Level of criticality	Vulnerability type
1.3: Curb lane parking for uncontrolled parked vehicles is placed unacceptably close to the building	Elevated	Primary
1.7: No private vehicle access control at the perimeter of the building	Elevated	Primary
2.5: Doors and walls along the security screening not adequately reinforced	Elevated	Secondary
2.7: Critical assets (people, activities, building systems and components) are not well separated from main entrance, vehicle circulation, parking	Elevated	Secondary
2.10: Ceiling, internal walls, overhead utilities and lighting systems are not designed to remain in place without generating debris in hazardous events	Extreme	Secondary
4.1: Low designed or estimated protection level of the building envelope against a possible high magnitude explosive threat	Extreme	Secondary
4.3: Glazing of the building are not secure in case of blast	Elevated	Secondary
4.4: Building is not designed to resist to high external pressure (ex. blast)	Elevated	Secondary
5.6: No redundant and reliable electrical service source	Extreme	Secondary
6.4: No provisions for air monitors or sensors for CBR agents	Extreme	Secondary
6.5: No method for fast air intakes and exhausts closure when necessary	Elevated	Secondary
7.11: No mass notification system that reaches all building occupants	Elevated	Secondary
8.1: CCTV cameras used, 24 hours/7 days a week recorded and monitored at the perimeter and in the critical areas of the building are insufficient	Elevated	Secondary
8.2: Video quality not adequate both in daylight and darkness	Elevated	Secondary
9.4: Emergency plan not up-to-date and not well designed	Elevated	Secondary
9.5: Operational continuity plan to apply no up-to-date and well-designed	Elevated	Secondary

5 DISCUSSION OF THE RESULTS

The case study analysed shows some interesting properties of the proposed BVAM method. It can be observed that:

- the adoption of the prototype BCA software tool developed for the analysis of the criticalities of the building greatly simplifies the activity of the AT. Furthermore, Table 7 provides, in a single screen, an effective description of the general criticalities of the building. It also provides a direct indication of the most significant areas where possible countermeasures for the mitigation of the vulnerabilities should be applied.
- the detailed description of the threats carried out in Step 2 highlights which criticalities are realistically exploitable, providing precise indications for the design of countermeasures.
- Step 3 allows for the selection of an appropriate VL by portraying a clear picture of what specific criticalities have emerged as a result of Step 2.

6 CONCLUSION

The proposed BVAM provides the assessment team with a qualitative and quantitative value assigned to the vulnerability of the building analysed, based on a vulnerability scale of seven levels. This value not only takes into account the physical and organisational aspects of the building, but also some of the social, economic, structural and institutional factors for different types of threats. The method described allows for the analysis of different kinds of vulnerabilities and the results obtained are useful for assessing the overall risk of different buildings for different threats. This enables for the prioritisation of actions and investments aimed at reducing vulnerabilities and thus reducing risk by enhancing the preparedness, protection and resilience of the buildings.

As a final consideration, it can be highlighted that the case study analysed shows consistent and easily interpretable results and objective assessments. This enables for the conduction of a coherent analysis and for the attainment of reliable results in an extremely complex context such as that related to risk assessment for terrorist attacks on a building.

REFERENCES

[1] Carbonelli, M., *Terrorist Attacks and Natural/Anthropic Disasters: Risk Analysis Methodologies for Supporting Security Decision Making Actors*, Aracne CBRN Series: Rome, 2019.

[2] Ayyub, B.M., *Risk Analysis in Engineering and Economics*, University of Maryland, Chapman and Hall/CRC, New York, pp. 35–38, 2003.

[3] Biringer, B.E., Matalucci, R.V. & O'Connor, S.L., *Security Risk Assessment and Management: A Professional Practice Guide for Protecting Buildings and Infrastructures*, John Wiley, 2007.

[4] Bouchon, S., *The Vulnerability of Interdependent Critical Infrastructures Systems: Epistemological and Conceptual State of-the-Art*, European Commission, Directorate-General Joint Research Center (JRC), Institute for the Protection and Security of the Citizens, Ispra, 2006.

[5] Modarres, M., *Risk Analysis in Engineering: Techniques, Tools and Trends*, Taylor and Francis: Boca Raton, FL, 2006.

[6] Sotic, A. & Radjic, R., The review of the definition of risk. *Online Journal of Applied Knowledge Management*, **3**, pp. 17–26, 2015. www.iiakm.org/ojakm/articles/2015/volume3_3/OJAKM_Volume3_3pp17-26.pdf.

[7] ISO 31010, Risk Management: Risk Assessment Techniques, International Organization for Standardization, 2009.
[8] European Commission Staff Working Paper, Risk Assessment and Mapping Guidelines for Disaster Management, Brussels, 2010. ec.europa.eu/echo/files/about/COMM_PDF_SEC_2010_1626_F_staff_working_document_en.pdf.
[9] DHS, National Infrastructure Protection Plan, Homeland Security Department, 2006. www.hsdl.org/?abstract&did=464612.
[10] UNISDR, Report of the Open-Ended Intergovernmental Expert Working Group on Indicators and Terminology Relating to Disaster Risk Reduction: Report of the Second Session (Informal and Formal). The United Nations Office for Disaster Risk Reduction, Geneva, Switzerland, 2016. www.preventionweb.net/files/50683_oiewgreportenglish.pdf.
[11] USA Federal Emergency Management Agency, *Reference Manual to Mitigate Potential Terrorist Attacks Against Buildings*, Fema 426/BIPS06, Edition 2, 2011. www.dhs.gov/xlibrary/assets/st/st-bips-06.pdf.
[12] Rouad, M., *Probability, Statistics and Estimation*, 1st ed., 2013, English version 2017. www.incertitudes.fr/book.pdf.
[13] Dusenberry, D.O., *Handbook for Blast Resistant Design of Buildings*, John Wiley, 2010.
[14] Karlos, V. & Larcher, M., *Guideline on Building Perimeter Protection: Design Recommendations or Enhanced Security Against Terrorist Attacks*, European Commission, Joint Research Center (JRC): Luxembourg 2020. op.europa.eu/en/publication-detail/-/publication/6d7e5311-f7c3-11ea-991b-01aa75ed71a1/language-en.

STRENGTHENING OF DISASTER RISK MANAGEMENT STRATEGIES IN THE PERUVIAN RAINFOREST IN THE FACE OF DEBRIS FLOW THROUGH A VULNERABILITY APPROACH

LUIS IZQUIERDO-HORNA, ANGELICA SÁNCHEZ-CASTRO & JOSE DURAN
Department of Engineering, Universidad Tecnológica del Perú, Peru

ABSTRACT

Disaster risk management (DRM) is a social process, which aims to prevent, reduce and periodically control disaster risk factors in society, taking into account both national and international policies, strategies and actions. In Latin America and the Caribbean, 25 countries aim to reduce risk, Peru is one of them with 1,535 provincial and local governments that have DRM instruments, however, in the Peruvian rainforest, there is a notable lack of these, meaning the population does not know what to do in an emergency situation. Of the multiple hazards to which Peru is exposed (i.e., earthquakes, mass movements, rainfall, etc.), 18% of these are debris, sludge and avalanche flows. Because of this, it is important to strengthen DRM strategies, specifically in the Peruvian rainforest, through an intercultural approach to reduce the level of social vulnerability to the debris flow. These strategies have to align with the Sendai Framework (2015–2030) and, at the same time, respond to the specific characteristics of each place due to the multicultural and multilingual nature of each sector. Therefore, this research proposes a holistic review for the improvement of vulnerability reduction strategies in social terms, with special emphasis on the intercultural dimension. In order to study the vulnerability factor, the conditions of exposure, fragility, and resilience were analysed, as well as urban centres, vital lines and services, poverty levels, socioeconomic fragility, and social organization levels. Finally, administrative decision-making, policy organization and implementation must respond to this cultural diversity and their capacities to cope with the adverse effects of disasters need to be strengthened.

Keywords: debris flow, disaster risk management, disaster risk strategies, vulnerability index, Peru.

1 INTRODUCTION

Strengthening disaster risk management (DRM) strategies in the Peruvian rainforest through an intercultural approach, regarding debris flow to reduce the level of social vulnerability, is a preponderant task considering the efforts that have been made in Latin America and the Caribbean during the last decades. In this way, the Risk Management Index for Latin America and the Caribbean (2020) has been developed, which allows understanding and evaluating the risk of disasters and humanitarian crises. In addition, it is also used for decision-making in prevention, preparation, monitoring, and response tasks. This information is relevant to determine, in the medium and long term, the effectiveness of the implemented strategies. This index is made up of the following dimensions: hazard and exposure, vulnerability, and lack of capacity [1]. The first dimension is determined by nature and cannot be reduced. The second is determined by the population and can be explained through multiple approaches (e.g., social, environmental, economic, etc.). Both dimensions interact with the lack of response capacity at the institutional and infrastructure level to make up this evaluation index that varies in a range from 0 to 10 in terms of risk [1]. Low and high values indicate a lower and higher level of risk, respectively. These indices facilitate the relative comparison of risk within the same country, as well as a comparison between countries. In this context, for the year 2020, Peru showed a level 6 according to this index, which is equivalent to a high risk [1].

WIT Transactions on The Built Environment, Vol 214, © 2022 WIT Press
www.witpress.com, ISSN 1743-3509 (on-line)
doi:10.2495/SSR220091

This research seeks to analyse the evolution process of strategies to reduce vulnerability due to debris flow implemented in the Peruvian rainforest in order to improve the conditions and livelihoods of the target population. Thus, in 2005, the Hyogo Framework for Action established a milestone in disaster reduction, whose general objective was to increase the resilience of nations and communities to disasters, as well as considerably reduce the loss of life and social, economic and environmental assets of communities and countries [2]. In this way, DRM came to occupy an important and priority place on the international agenda, in which governments had the responsibility to deal with disaster prevention, risk management, and vulnerabilities [2]. By 2015, progress was made thanks to its application, with advances in disaster risk reduction observed at the local, regional, national and global levels, thus demonstrating that DRM is a profitable investment in preventing future losses [2]. In this sense, considering the constant growth of disaster risk, the increase in the degree of exposure of people and goods, in combination with the lessons learned from past disasters, the need to strengthen disaster preparedness becomes clear. It is for this reason that it is necessary to adopt early measures to ensure an effective response and recovery at different levels of risk [3].

International mechanisms have contributed fundamentally to the development of risk reduction policies and strategies, whose corrective and prospective approach is centred on people. To direct these policies, governments must lead, regulate, coordinate DRM and interact directly with citizens who must be taken into account in their design and application in order to integrate disaster risk into their management practices [4]. Therefore, in 2014, Frontline Vision (VPL) through the project "Perspectives on the implementation of the Hyogo Framework for Action (MAH) in South America" sought to know the perception of the population vulnerable to disasters, in relation to DRM policies and their execution at the local level. To do this, they created spaces for dialogue and meetings between citizens, authorities and organizations present in the area. They concluded that, in recent years, the governments of Latin America have made significant progress in the institutional framework, design and implementation of national DRM policies. Parallel to this, there is also evidence of a weak institutional framework at the local level and the need to articulate the actions of the central government with local governments [5].

In the case of Peru, in 2010, the National Policy: Disaster Risk Management (PLANAGERD 2014–2021) was implemented, and in 2011 the regulatory framework was approved, which makes viable decentralized actions at the local level, under the protection of the Law No. 29664 and its regulations, transferring responsibility for action to local governments. For development planning and management, there is the National Disaster Risk Management System (SINAGERD) and for corrective management, there is Civil Defense, whose strategic objective is focused on integrating disaster risk into policies, plans, and sustainable development programs at all levels. A DRM capacity development policy was also promoted; however, there are still constant limitations regarding the allocation of financial resources and/or operational capacities. In this sense, knowledge on disaster risk reduction is made available to everyone, especially the most vulnerable and poor communities, to achieve the construction of resilient communities. In short, an institutional commitment has been achieved, but the achievements are not extensive or noticeable [6]. Along the same lines, in 2015 the Sendai Framework was implemented, which aims to strengthen disaster risk governance, considering regional, national and international platforms. One of its seven goals is to reduce the damage caused by disasters in vital infrastructure and the interruption of basic services, such as health and education facilities, building their resilience by 2030 [3].

In Peru, during 2019, 1,872 municipalities complied with reporting on DRM, of which approximately 80% reported that they have a DRM Unit or Civil Defense Office. From this information it was observed that 6% corresponds to the Junin region (region of interest belonging to the central rainforest of Peru). On the one hand, 1,428 municipalities reported that the Civil Defense office carried out prevention actions in the population and that 5.7% of these are from Junin. On the other hand, at the national level, 74 local governments did not carry out any action and three of them belong to Junin. Likewise, 736 took training courses in reactive risk management, of which 30 municipalities are from Junin. As far as drills are concerned, 1,236 executed drills in a coordinated manner with different institutions and evaluated the local emergency operations plan where 74 are from Junin. In relation to identifying the level of risk, 793 fulfilled this task and of these 43 are from Junin. Supervising compliance with safety standards in venues with public access was carried out by 572 municipalities, of which 26 belong to Junin. Finally, 684 properly marked security zones in high-risk areas and 28 belong to Junin [7]. That said, by 2020, the Junin region had nine provincial governments, 109 district governments, and 145 small communities. Of these, 102 municipalities have risk management instruments. The problem lies in those municipalities that do not comply and do not have these instruments, so the population does not know how to respond to the dangers, risks and vulnerabilities when disaster situations occur [8]. It is important to emphasize that Junin is one of the regions that is most affected by landslides during the winter seasons, such is the case of the María Pía streams that affects Pampa del Carmen settlement, which is the case study considered for the development of this research.

It can be deduced, then, that the purpose of this evaluation is to strengthen the current strategies to reduce the level of vulnerability based on the gaps evidenced as a result of recent events (e.g., linguistic, economic, origin, resilience, etc.) raised in Peruvian territory. Finally, this study proposes some recommendations that will help a more efficient management of resources and people.

2 BACKGROUND AND CASE STUDY INFORMATION

In recent years, Peru has recurrently suffered the impacts generated by the occurrence of debris flow, especially in the northern and central regions, due to the fact that it is located on medium and high slopes, rivers and streams [9]. Debris flow emergency records are approximately 1,529 from 2003 to 2018. Fig. 1 shows the frequencies recorded during that time interval. It can be seen that during 2017, the highest record was 570 incidents and the lowest record was in 2011 with 45 events [10]. Likewise, the impact generated in society is 439,636 people affected and 5,243 homes destroyed, with 2017 being the year with the most reported cases, registering 115,431 people affected and 2,365 homes destroyed [10].

2.1 General characteristics of the Junin region: Central rainforest

The exploratory analysis of this research is based on the Chanchamayo province of the Junin region, located in the central rainforest of Peru, which is geographically located at an altitude of 775 m above sea level, with a south latitude of 11°03'21" and west longitude of 75°19'45" [11]. It has temperate and humid climates with temperatures ranging between 23°C to 32°C [12]. Junin has a population of 1,246,038 people and the district of Chanchamayo has 27,790 people, approximately [13]. Regarding the economic sector, 41.8% of the population is dedicated to agriculture and fishing, with mining being the economic activities with the lowest active population with 2.2% [11]. Likewise, in relation to access to basic services, certain gaps are evident where 7.2% of the population does not have potable resources, 35.7%

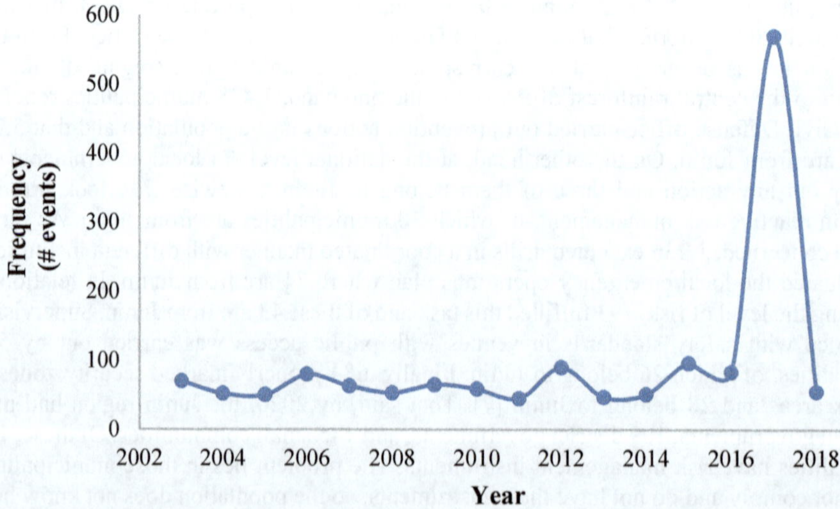

Figure 1: Occurrence of debris flows in Peru during the last 20 years.

does not have hygienic services, 8.4% does not have electricity and, approximately, 2.2% of this population does not have household garbage collection services, among other deficiencies in the management of their capacities [14]. In addition, there are sectors located near streams where accessibility to basic services is exponentially difficult [15].

2.2 Representative events that occurred in Peru (2000–2021)

2.2.1 La Merced

In the Pampa del Carmen sector, La Merced district, on 23 August 2003, a debris flow originated due to heavy rainfall that caused the activation of the María Pía stream, affecting 300 homes, 200 m of road, among other infrastructures [16]. In the same way, on 22 January 2007 there were rains for 9 hours that reached up to 173.7 mm, causing the streams to produce debris flows [17], thus generating 1,758 people affected, 283 houses destroyed, 213 houses affected [18]. On the other hand, considering that the area is influenced by tropical, humid and rainy weather that cause massive landslides [19], on 21 February 2021, a landslide occurred in the district of La Merced, province of Chanchamayo [20]. As a result of this incident, it was reported that 60% of the population (i.e., 250 people affected) were left without drinking water service, 27 houses destroyed, 108 houses affected, causing 83 homeless people [21]. Additionally, that same day the authorities carried out a damage assessment and moved heavy machinery for debris removal and cleaning work. Also, the municipal government personnel and the regional government provided help to the affected families.

2.2.2 Pataz

In the district of Parcoy, province of Pataz, on 12 April 2009, the first case of landslide was recorded near a stream due to heavy rains, land with steep slopes, and increasing deforestation in the area. The result was nine deaths, 150 people affected and five houses destroyed [22].

Likewise, on 15 March 2022, a debris flow slide occurred due to increased rainfall and poor housing distribution in a steep area. There was a mass movement that caused damage to life, health and housing, where there were 53 victims, eight destroyed homes, 12 uninhabitable homes, 13 affected homes, a business establishment destroyed and 100 m of adduction line destroyed [23]. In addition, on that day, the authorities carried out their first search 11 hours after the landslide occurred with the help of the National Police. On the other hand, the Municipality of Parcoy moved heavy machinery and the Regional Government provided search support and specialized rescuers. On 22 March, the MINSA provided psychological help to 12 residents for the loss of their relatives and implemented a refuge area [23].

2.2.3 Santa Ana

On 16 February 2021, intense rainfall occurred that caused a mudslide, damaging homes, institutions, crops, roads and, above all, life and health in the city of Quillabamba and the Chaupimayo B sector, in the district of Santa Ana, province of La Convencion, in the department of Cusco [24]. This was due to the fact that the Chaupimayo stream was activated as a result of the rains, causing landslides where the flow travelled around 7.8 km. Likewise, this fact caused the death of four people and the disappearance of five people. Also, ten homes were affected, one educational institution destroyed, one educational institution uninhabitable, 0.3 km of highway affected and two pedestrian bridges destroyed [24]. After this incident, the authorities provided aid to the affected people, where food was offered to the victims. A temporary shelter was also installed and, later, heavy machinery was assigned for clean-up work [24].

Fig. 2 shows the main ravages of the events described above, while Table 1 identifies the main characteristics of the Junin region in the central Peruvian rainforest. These social, economic and environmental characteristics serve as a reference to establish a possible baseline for the design, planning and implementation of strategies to reduce vulnerabilities when facing debris flow.

3 ANALYSIS OF THE IMPLEMENTATION OF STRATEGIES TO REDUCE VULNERABILITIES AGAINST DEBRIS FLOW

The application and evolution of strategies to reduce vulnerability due to debris flow implemented in the central rainforest in order to improve the conditions and livelihoods of the population from 2005 to 2022, leads us to analyse the policies, plans and strategies within the Hyogo Framework for Action 2005–2015, which is based on the strategic and systematic approach to disaster vulnerability reduction [2] and the Sendai Framework for Disaster Risk Reduction 2015–2030, which aims to prevent new and existing risks by integrating economic, structural, legal, cultural, educational, environmental and technological measures [3]. Both are DRM instruments to be applied in all UN member countries in order to build the resilience of nations and communities to disasters [3].

3.1 Analysis of DRM strategies in the face of the PLANAGERD incursion

In Peru, there are sectors where poverty and extreme poverty coexist, which, along with their low resilience, magnify the impact generated by debris flows on the population and housing, making evident the need to improve DRM strategies [26]. In 2012, the Ministry of the Environment warned –through the physical vulnerability map– that 46% of the national territory is in conditions of high to very high vulnerability [27] and that 36.2% of the national population is using and occupying this space in order to take preventive actions and safeguard

Figure 2: Evidence of destruction due to debris flow in La Merced, Pataz and Santa Ana, respectively [20], [23], [24].

Table 1: Social, economic and environmental characteristics of the sector of interest.

Dimension	Description in the sector	Source
Social	• There are gaps in the services of drinking water, electricity and hygienic services • Excess of family members residing in a dwelling • Low percentage of professionals	[11] [25]
Economic	• Income is mostly based on agriculture • They have houses with predominant material of bricks, adobe, quincha, etc. • The largest economically active population is engaged in agricultural work	[11] [19]
Environmental	• Presents humid climate with high rainfall • Temperatures around 32°C • Low resistance soils • Location close to streams	[12]

the integrity of the occupants [26]. For 2013, in the Junin region, the number of vulnerable population rose to 810,236. For 2016, 810,770, and for 2021, 896,422. These data allow us to understand the need to incorporate the PLANAGERD 2014–2021 as a policy instrument to apply DRM strategies aligned with the strategic objectives of the MAH 2005–2015 and State policies No. 32 "Disaster Risk Management" and No. 34 "Territorial Planning and Management" [26].

Effective disaster risk management is conducive to sustainable development. Countries that have implemented such strategies have improved their disaster risk management capacities. Overall, the Hyogo Framework for Action has been an important instrument for raising public and institutional awareness to generate political commitment and to focus and galvanize action by a wide range of actors at all levels [26]. Likewise, oriented to the strategic processes regulated in Law No. 29664, which are: estimation, prevention, reduction, preparation, response, rehabilitation, reconstruction, institutionality and culture of prevention as DRM strategies [28]. Thus, in order to incorporate an efficient DRM and achieve a safe and resilient society, the different levels of government make use of the following instruments: (a) The Policy and the PLANAGERD, which establish the strategic lines, the objectives and the necessary actions to implement the SINAGERD Law; (b) The financial management strategy for disaster risk by the Ministry of Economy and Finance, whose purpose is to ensure adequate financial capacity in DRM processes; (c) The mechanisms for coordination, decision, communication and information management, which promote inclusive spaces where it is sought that public and private entities and organized civil society participate, coordinate, articulate and integrate proposals to raise awareness and generate political and institutionalize disaster risk prevention and reduction; (d) the National Information System for Disaster Risk Management; and (e) the National Radio for Civil and Environment Defense (Communication system for DRM) [28].

Within the strategic objectives of PLANAGERD, it is pointed out that it is essential to strengthen the participation of the population and organized society for the development of a culture of prevention. For this, it is established to promote the incorporation of DRM in basic and higher education, under the responsibility of regional and local governments, universities, social organizations and private entities; develop community education programs in DRM aimed at the urban and rural population, incorporating the rights approach and interculturality, and finally promote good practices in DRM in the urban and rural population, respecting cultural diversity and involving the media [26].

3.2 Capacity: Response relationship with the sustainable development goals

Evidence of the evolution that has occurred in the Peruvian rainforest is that, in 2020, 75.3% of regional and local governments organized working groups for DRM made up of officials and authorities, led by the highest executive authority. Of this, 96.9% installed the work group, 95.1% appointed the technical secretary of the work group, 65.6% have an annual work plan and 56.6% have internal regulations for functioning. With these data, we can conclude that one of the priorities of the Sendai Framework is being worked on: strengthening disaster risk governance. However, there is still a gap to close with 24.7% of municipalities that reported not having constituted DRM working groups [29].

Regarding investing in disaster risk reduction, according to the Sendai Framework, it is materialized in our country in the Results-Based Budgeting Program No. 068 "Reduction of Vulnerability and Attention to Disaster Emergencies" [30]. This program has the specific result of reducing the vulnerability of the population and their livelihoods in the event of hazards. In 2019, 58% of provincial and district municipalities carried out activities and/or investment projects for this purpose. Of this total, 59.1% carried out actions aimed at avoiding the generation of new risks, 50.3% carried out actions to reduce vulnerability and existing risks, 43.2% developed actions in order to seek, execute and rehabilitate the areas affected by a disaster, 26.2% estimated the risk levels and 21.7% carried out actions for reconstruction [29].

4 DISCUSSION

Considering that the strengthening of the strategies arises as a product of the reflexive and practical analysis of the differences of each one of the peoples [31], the establishment of DRM mechanisms and strategies evidenced in recent events for the recovery, rehabilitation and reconstruction phase are a good sign of progress. However, there are still social gaps that increase the negative impacts. We believe that the strengthening of these strategies should be carried out taking into account the characteristics of the case study and promoting citizen participation. In short, recovery processes should take into account the characteristics of the territory, lifestyles, and respect for traditions and customs. On the other hand, economic and social rehabilitation and recovery would only be achieved with the participation of the beneficiaries and it would be key to know this human group and its dynamics in order to formulate and apply the corresponding policies.

On the other hand, in the process of strengthening strategies, the National Policy for Disaster Risk Management to 2050, approved by DS. No. 038-2021-PCM, suggests that meeting its strategic objectives will depend mainly on the establishment of tools that allow universal access to information and knowledge available on DRM [32]. Thus, according to the analysis conducted on Municipal Management between 2008 and 2014, there was an 8.8% increase in the number of municipalities that reported having technical instruments for Civil Defense [33]. With the gradual implementation of these instruments, the region is more aware of the level of exposure to debris flows, so much so that in 2010, the National Water Authority, through the Local Water Authority, launched the prevention plan for natural phenomena such as floods, landslides, landslides and droughts, thus promoting a series of structural measures in the sector (e.g., construction of 200 ml of gabions and reforestation of 1 ha) [34]. Similarly, in 2015, there was an increase of 9.8% in the number of municipalities that reported having disaster risk management instruments [35] and the DRM working group of the Municipality of Chanchamayo was formed in compliance with Law No. 29664 and its Regulations [36]. Finally, in the following years, the level of awareness increased [37], leading to the implementation of related instruments in the pending municipalities [38].

In regulations terms, in 2021 the Plan for Prevention and Risk Reduction in the face of the rainy season in the district of Chanchamayo 2021–2022 was approved, the purpose of which was to carry out intervention activities, since in recent years' natural disasters have occurred, with heavy damage and losses caused by heavy rains, landslides, rock falls, flooding in different sectors of the population due to overflowing rivers. This plan seeks to guarantee and implement activities for the clearing and channelling of rivers and drainage of rainwater channels in the province of Chanchamayo [39].

5 CONCLUSION

To summarize, DRM as a social process has the purpose of periodically preventing, reducing and controlling disaster risk factors by applying national and international policies, strategies and actions. For this, provincial and local governments need to have DRM instruments, mainly those located in the Peruvian rainforest. And for this reason, being a multilingual and multicultural society, these strategies require respect for an intercultural approach to reduce the level of social vulnerability, mainly in the face of debris flow. Likewise, this new approach would be complemented by the current strategies applied in central, provincial and local governments that are aligned with MAH 2005–2015 and the Sendai Framework (2015–2030). Within this intercultural approach suggested to strengthen DRM strategies, the cultural elements that are feasible to identify from the daily practices of the inhabitants should be considered, since they would constitute pillars, going from being a generic instrumentation to a risk social management, also a management that responds to the characteristics of the territory and society. Key aspects such as language – which was often an obstacle in communication processes – will be especially relevant when you want to convey guidelines, ideas or proposals. Respect for territory and culture will facilitate the implementation of DRM strategies. In this sense, a coordinated and collaborative work will be possible to carry out for the improvement of conditions and livelihoods of the affected population.

ACKNOWLEDGEMENTS

This project was funded by Universidad Tecnológica del Perú, within the framework of the "Research Projects I+D+i 2021 – 2" agreement.

REFERENCES

[1] INFORM, Risk Management Index for Latin America and the Caribbean. pp. 1–30, 2020.
[2] ONU, World Conference on Disaster Reduction Report. **72**, pp. 1–44, 2005.
[3] ONU, Sendai Framework for Disaster Risk Reduction 2015–2030. pp. 1–40, 2015.
[4] EIRD, National report on the progress in the implementation of the hyogo framework for action, Peru. pp. 1–18, 2007.
[5] Carbonel, D. et al., A local, critical and provocative perspective on the implementation of risk management policies in South America. pp. 1–74, 2015.
[6] CENEPRED, Informe Nacional del Progreso en la Implementación del Marco de Acción de Hyogo (2011–2013). pp. 1–49, 2013.
[7] CENEPRED, Municipalities that reported having a disaster risk management unit or civil office, with media. p. 1, 2019.
[8] CENEPRED, Municipalities with disaster risk management (GRD) instruments, by department, 2020. p. 1, 2020.
[9] Fidel, L. et al., Map of susceptibility to mass movements in Peru. **9**, pp. 308–311, 2010.
[10] INDECI, INDECI 2019 statistical compendium on GRD preparedness, response and rehabilitation. pp. 1–122, 2019.
[11] INEI, Junin statistical compendium 2017. pp. 1–639, 2017.
[12] INGEMMET, Critical geological hazard zones in the Junín region. pp. 1–50, 2014.
[13] INEI, Final results. **1**, pp. 1–1069, 2018.
[14] INEI, Access to basic services in peru 2013–2018. pp. 1–52, 2019.
[15] INGEMMET, Geological hazard assessment for mass movements in Alto Capelo, La Cruz, San Carlos, María Pía, Abanico and Potoque sectors. pp. 1–56, 2021.
[16] Medina, L. & Núñez, S., Debris flow that occurred on January 21, 2007 in the San Ramón locality. pp. 1–6, 2007.
[17] INGEMMET, Geologic hazards on January 21 in San Ramon district. pp. 1–38, 2007.
[18] PREDES, Preliminary report emergency in San Ramon: Chanchamayo, Junín. pp. 1–9, 2007.
[19] MPCH, Report No. 229-2021/GOP/MPCH. pp. 1–121, 2021.
[20] INDECI, Landslide in the district of La Merced, Junin. p. 4, 2021.
[21] EDAM, Peru Form. pp. 1–5, 2021.

[22] INGEMMET, Technical inspection of geological hazards due to mass movements Sanchez Carrión and Pataz provinces, La Libertad region: Chamanacucho, Tayapampa, Retamas, Pataz, Collona and Sartimbamba sectors. pp. 1–48, 2009.
[23] INDECI, Landslide in the district of Parcoy: La Libertad. pp. 1–33, 2022.
[24] INDECI, Landslide in the district of Santa Ana: Cusco. pp. 1–18, 2021.
[25] INEI, Final results. **1**, pp. 1–865, 2018.
[26] SINAGERD, National disaster risk management plan 2014–2021. pp. 1–63, 2014.
[27] MINAM, Physical vulnerability map of Peru. pp. 1–67, 2011.
[28] CENEPRED, Law of the national disaster risk management system, SINAGERD law No. 29664. pp. 22–26, 2014.
[29] INEI, Peru: Municipal management indicators 2020. pp. 1–236, 2020.
[30] PCM, Vulnerability reduction and disaster emergency response budget program. pp. 1–946, 2020.
[31] Aid C, Practicas S. Intercultural risk management: a local experience. pp. 1–78, 2014.
[32] PCM, Supreme Decree No. 038-2021 The National Policy on Disaster Risk Management to 2050. pp. 1–72, 2021.
[33] INEI, Peru: Municipal management statistics 2008–2014. pp. 1–395, 2015.
[34] MINAGRI, Prevention plan for natural phenomena such as floods, landslides, landslides and droughts. pp. 1–136, 2010.
[35] INEI, Peru: Municipal management statistics 2011–2015. **1**, pp. 1–390, 2016.
[36] MPCH, Mayoral Resolution No. 044-2015 ALC/MPCH. pp. 1–2, 2015.
[37] INEI, Peru: Municipal management indicators 2017. pp. 1–329, 2017.
[38] INEI, Peru: Municipal management statistics 2019. pp. 1–278, 2019.
[39] MPCH, Mayoral Resolution No. 477-2021 ALC/MPCH. pp. 1–4, 2021.

BEST PRACTICES FOR VULNERABILITY MANAGEMENT IN LARGE ENTERPRISES: A CRITICAL VIEW ON THE COMMON VULNERABILITY SCORING SYSTEM

JAQUELINE HANS & ROMAN BRANDTWEINER
Vienna University of Economics and Business, Austria

ABSTRACT

Over the past decade, enterprises have been increasingly suffering from attacks conducted by cybercriminals. Potential losses are not only reflected on their revenue or stolen data, but also on their damaged reputation. Most often, these attacks were possible due to the successful exploitation of vulnerabilities within the company's system. Many of such attacks could have been mitigated, if responsible actors took the right actions related to the management of such vulnerabilities. This paper aims to summarize good practices regarding vulnerability management, with essential focus on the matter of prioritization. For this, several vulnerability scoring systems such as the Common Vulnerability Scoring System were analyzed according to the way they are portrayed in scientific literature. It will also analyze non-technical, human factors as well by reflecting on organizational aspects. The aim is to provide an overview about the options large enterprises have in this regard and to inform about potential consequences they could face. It will also reflect on the problematic behind the trade-off between investing enough in a cybersecurity foundation, while simultaneously remaining profitable.

Keywords: cybersecurity, e-security, vulnerability scoring system, CVSS, vulnerability management.

1 INTRODUCTION

Due to the increasing digital reliance, enterprises are constantly exposed to the vulnerabilities of their systems. Especially large corporations encounter an incredible number of vulnerabilities on a daily basis and, at some point, struggle to categorize these vulnerabilities prioritize the elimination of those. Automated processes can be a very efficient way to quickly assess the severity of each vulnerability, but most approaches still lack accuracy and trustworthiness. Especially the Common Vulnerability Scoring System (CVSS), is a widely used and accepted standard [1]. However, there is a large disagreement within cybersecurity experts whether using this standard alone should be considered a good practice or even is reliable at all [2]. This arises the question whether and how the CVSS could be improved, if it should be complemented by alternative practices, or if it should not be used at all. Furthermore, it should be reflected on whether vulnerability management is a question of technological kind alone or whether there are several human and organizational factors influencing this matter as well [3].

This literature synthesis aims to provide a summary that does not focus on certain algorithms and methodologies alone, but aims to create a big picture, cybersecurity researchers and practitioners can build upon and derive implications from. In the first section, the context of vulnerability management in large enterprises is explained by emphasizing the consequences of vulnerability exploitations and analysing this topic on the surface. The following section deals with vulnerability prioritization and scoring systems such as CVSS by reflecting on its scope and responsibilities, and by comparing it to different models. Lastly, the final section deals with complementary or alternative approaches and viewpoints which do not necessarily relate to severity scoring alone but might be a good integration to vulnerability management processes in general.

WIT Transactions on The Built Environment, Vol 214, © 2022 WIT Press
www.witpress.com, ISSN 1743-3509 (on-line)
doi:10.2495/SSR220101

2 METHODOLOGY AND RESEARCH DESIGN

The methodology applied for this study is a systematic literature review. Hence, the foundation of this research is primary literature found in several digital databases. The aim is to include as many findings as possible in order to obtain a large number of insights regarding vulnerability management in enterprises by including the perspective of cybersecurity researchers as well.

2.1 Quality criteria

The selection of the primary literature is based on several quality criteria: The first criterium is accessibility, which means that a full-text version of each source should at least be available within the WU library network. WU stands for "Wirtschaftsuniversität" (University of Economics and Business) that is the name of our university, i.e. we used this specific library network for our research. The second criterium is that all sources need to be peer-reviewed as this ensures an acceptable amount of trustworthiness and correctness of information. Also acceptable are resources that were published by official sources such as FIRST.org, the creators of the CVSS, Tripwire, OWASP, and NIST. Lastly, the final criterium is relevance. That means that the primary literature must be relevant according to the topic it addresses, but also according to the date it was published.

2.2 Database search and literature screening

The WU library network, which is our main source for finding primary literature, refers to the WU library catalogue and the WU library cataloguePLUS including other digital databases that allow full-text access such as Springer, IEEE, EBSCO, ProQuest, Wiley Online Library, and ScienceDirect. Within these databases, resources were filtered by the following keywords.

General keywords: Vulnerability management, threat management, system vulnerability, network vulnerability, cybersecurity risk management, vulnerability assessment, vulnerability prioritization, patch management, vulnerability severity scoring.

Specific keywords: CVSS, Common Vulnerability Scoring System, FIRST, base metric, temporal metric, environmental metric, Tripwire IP360, Tripwire Vulnerability Scoring System, OWASP Risk Rating Methodology, OWASP Vulnerability Scoring.

The search results based on the aforementioned keywords were further screened based on title, abstract, content, and selected under fulfilment of the previously defined quality criteria: 134 elements were selected, based on title of the paper from these 134 articles 59 were selected based on the content of the abstract and out of them 11 elements were selected based on content and quality criteria. Those 11 articles were the final sample of the study.

3 RESEARCH AIM AND OBJECTIVES

The aim of this systematic literature review is to provide new insights regarding best practices of vulnerability management in enterprises as well as to reflect on the CVSS by analysing as much primary literature as possible. The final deliverable should be a summary that not only highlights the key findings but also the implications they have for future research and practice. Hence, the paper further aims to identify potential research gaps as well.

While there are many papers that introduce certain models for vulnerability prioritization or reflect on certain frameworks alone, the subject of vulnerability management is lacking research contributions that summarize, compare, and reflect on them simultaneously and, hence, provide an overview of many existing, yet unknown solutions. This paper summarizes good practices, reflects on widely used standards, and aims to introduce new ways of thinking about vulnerability management by paying not only attention to the technical details but also taking financial and organizational factors into consideration. This paper is guided by three specific research questions (RQ):

- RQ1: What is the scope of vulnerability management and what are contemporary issues and consequences enterprises face?
- RQ2: Is the widely accepted CVSS accurate, trustworthy, and reliable? How could it be improved?
- RQ3: What other factors and methodologies should be considered when it comes to vulnerability prioritization? Are there any alternative practices that could or should complement vulnerability severity scoring systems?

4 THE CONTEXT OF VULNERABILITY MANAGEMENT IN ENTERPRISES

In order to dive in into the importance of vulnerability management in enterprises, it is important to obtain an understanding of how the term "vulnerability" is defined. According to Sukaina Bhawani, a senior researcher at the Stockholm Environment Institute, the term vulnerability is referred to as "capacity to be wounded, i.e., the degree to which a system is likely to experience harm due to exposure to a hazard" [4].

While the definition's origin lies within the fields of geography and natural hazards, the concept of it has become applicable for other areas of research as well. Due to the wide range of different disciplines, several definitions have been established based on precisely determined factors such as key attributes, exposure units, and decision scale. Key attributes can be the capacity, sensitivity, and the exposure of such systems. The exposure units relate to the units that are affected by a potential exploitation. This can be individuals, groups of people, but also systems and other units. The decision scale, in its basic definition, mostly refers to the regional scope an exploitation would have [4].

A definition by Haimes, that relates more to the risks infrastructures and nations are exposed to, states that vulnerability interacts with other parameters such as intent, capability, threat, and risk. He underlines that risk modelling needs an integration of these parameters in order to understand the targeted infrastructure based on internal factors such as the current system's state and external factors such as the patterns of criminals selecting their targets [5].

This primarily refers to vulnerabilities/risks of nations' infrastructures. However, it is crucial to understand the similarities and differences between general vulnerability definitions such as the one by Bhawani, and more narrowed-down definitions such as the one of Haimes. This creates a broader understanding about the context of vulnerabilities that enterprises are exposed to due to their IT infrastructure.

Considering that this paper aims to primarily address the term vulnerability in the field of cybersecurity, this term will mostly refer to the exploitability of certain parts of IT systems and networks. However, it is important to recognize that these vulnerabilities also come along with several other consequences for companies, resulting in additional vulnerabilities with other key attributes, exposure units, and decision scales. A successful cyber attack does not only harm IT systems or networks, but it could also potentially harm the company and its stakeholders as well. Furthermore, victims could experience losses in terms of their data, their reputation, or their revenue.

Data that was gathered by Statista, and published by Accenture reveals the consequences of cyber-attacks global companies had to suffer from in 2018. According to the study, companies suffered from an annual average loss of US$5.9 million caused by information loss, US$4 million caused by business disruption, US$2.6 million caused by equipment damages, and US$0.5 million of revenue loss [6]. Another survey, that collected data from 522 cybersecurity practitioners, reported an average loss of US$288,618 caused by the successful exploitation of computer security vulnerabilities [3]. Considering the exponential growth of digital reliance and IT infrastructures, these numbers can be expected to significantly rise in the upcoming years.

The context of vulnerability management in institutions is further analyzed in a systematic literature review conducted by Uddin et al. [7]. During their analysis of various primary literature sources, they were able to discover essential research gaps and implications. The first research gap they discovered is the lack of knowledge regarding the relationship between operating costs and the exploitation of vulnerabilities. While most institutions understand that it is important to invest a certain amount of money regularly in their cybersecurity infrastructure, they sometimes lack information regarding a budget that is sufficient to achieve the best possible protection and at the same time does not immensely affect the financial return that most managerial decision making is based on. Although there are several policies and guidelines that determine a few of these decisions upfront, they are still considered being too ad-hoc and not fully reliable [7]. Nonetheless, the issue about an appropriate budget remains highly important. According to a pervious study, a lack of funding is, among other organizational factors, a primary non-technical cause of vulnerability exploitation [3].

As can be drawn from these findings, best practices regarding vulnerability management are not easily determined. This is due to enterprises' lack of understanding regarding the vulnerabilities within their systems and the meaning of the term "vulnerability" and its consequences in general. Furthermore, policies and guidelines are not precise enough frameworks to provide them with clear instructions regarding budgeting decisions as well as methods for appropriate vulnerability assessments. While the whole subject itself involves several research gaps and requires various policy adjustments, the following section of this paper aims to look at vulnerability management more precisely, paying essential detail to the prioritization of vulnerabilities by comparing several standards and practices.

5 THE CONTEXT OF VULNERABILITY MANAGEMENT IN ENTERPRISES

The following section will give an outline about the partly automized processes of vulnerability severity scoring. The aim is to define the scope of severity scoring systems by also reflecting on their performance, risks, and responsibilities.

5.1 CVSS

The most frequently used system of vulnerability severity scoring is the CVSS. Despite being criticized to some extent, CVSS has become an accepted standard when it comes to the scoring of vulnerability severities. Within the US National Vulnerability Database, all vulnerabilities are scored based on CVSS and also in many more areas, it is visible that this system seems to be the predominant methodology for vulnerability scoring [2].

The CVSS has been originated by FIRST, an organization that aims to improve incident response and other Internet security related operations. In their mission statement, the organization emphasizes its global position and reach around the globe [8].

Based on the official documentation, it can be noted that the CVSS' primary aim is to assess a vulnerability's severity and reflect on it based on a numerical score it produces. This score reaches from 0 to 10, where 0 represents a vulnerability of very low severity, and 10 represents a vulnerability of high severity and criticality. What should be additionally noted is that the score assigned to each vulnerability is relative to other vulnerabilities' severities

[9]. By this, the CVSS aims to support enterprises and other groups of interest with the prioritization of such vulnerabilities. Considering that the FIRST organization is constantly evolving and improving this severity scoring system, three main versions of it have been established over the years. This paper will, hence, mostly refer to the current version of CVSS, being CVSS v.3.1.

While this paper does not intent to analyse the mechanisms behind the CVSS in close detail, it is important to achieve a basic understanding of what it does and essentially to give an overview about its aim, purpose, and use case.

Simplified, the CVSS is composed of three main metric groups. The first metric group, the base metric group, contains metrics such as the attack vector, privileges required, the scope, or its impact regarding confidentiality, integrity, and availability. Within the base metric group there are also two subgroups, one referring to exploitability and the other one referring to the impact. While the other metric groups are considered optional groups, the base metric group is essential for determining the final severity score as it is constant over time and across user environments. The temporal metric group, on the other hand, does change over time, but not across user environments. Lastly, the environmental metric group refers to metrics that are tailored to the specific user's environment [9]. Summarizing, it can be said that the base metric provides an objective way of assessing vulnerability severities relative to other vulnerabilities, while the optional metric groups enable the user to adjust the scoring based on its unique characteristics.

Capturing these facts is very important when reflecting on some of the critique that CVSS is exposed to. The CVSS has been partly criticized for not sufficiently reflecting on individual and unique characteristics of vulnerabilities that are varying for each organization [2].

While it can be debated on whether the two optional metric groups are portraying these characteristics accurately, it should be noted that during the evolvement of CVSS, the FIRST organization has included additional metrics and accordingly adjusted them to make the assessment of vulnerabilities more individual and, therefore, the scoring more reliable and tailored to the IT infrastructure of the according environment. Hence, it can be assumed that FIRST is aware of potentially missing characteristics and is steadily improving, taking criticism by researchers and practitioners into consideration.

Furthermore, it has to be noted that by the time this paper is written, FIRST has distanced themselves from being entirely responsible for the vulnerability management processes of an organization. On their website, they state that factors such as risks regarding monetary losses or customer being affected by a breach go beyond the scope of CVSS and, therefore, CVSS can only give useful inputs and should not be used alone, but rather as a complementary methodology. The CVSS aims to measure severity and not risk and FIRST acknowledges that CVSS is only an addition to an existing, contextual risk assessment of an IT environment [10].

Reflecting once again on the different metric groups, FIRST suggests that the base score and the temporal score should be conducted by the assigned security professional/analyst, whereas the environmental score is determined by the end-user, for instance, a system administrator. However, while it is good that these various metrics and metric groups exist, they seem to be not much used in practice [2]. In a different study, the assessment difficulty of CVSS' environmental metrics was further analysed. In this controlled experiment that was conducted with the participation of 29 MSc students, the results revealed that for different states and variations of the network layout, the severity assessment using CVSS gets increasingly difficult. This also applies for the configuration of the network, indicating that on a system level, the correctness of the scoring might be impacted [11].

Furthermore, critics claim that the CVSS base score is still relatively unexplored and that its accuracy, whether it be due to overlooked factors or errors in the mathematical formula, is not yet fully proven. Nonetheless, determining the actual severity compared to the severity assigned by CVSS or other systems is a difficult matter. Previous studies tried to improve it using various approaches. Some approaches would additionally measure the time it would take to exploit the vulnerabilities, or by specifically looking at those vulnerabilities that are exploited in practice. These approaches, however, could also be seen as flawed due to many factors that are involved and that do not necessary indicate a higher severity by default [2].

Another method of analysing this subject is, hence, the conduction of expert interviews in which cybersecurity professionals estimate the severity for some vulnerabilities which is then compared to the severity score indicated by the CVSS. The results of one study revealed that some experts indeed assign different values to the vulnerabilities than the CVSS base score. It even claims, that these size of the variance goes beyond a level that actual users of the CVSS would feel comfortable with. According to this study, especially XXS (cross-site-scripting) vulnerabilities received a comparatively too low base score, and SQL injection vulnerabilities received a relatively high base score [2]. However, it should be noted that this particular study was carried out a while back and examined the CVSS v2 and, therefore, could possibly deliver different results if it was conducted nowadays, considering that the CVSS is steadily evolving and improving.

Another, more recent study conducted a survey with students and professionals, estimating vulnerabilities' severity using the methodology by CVSS v.3. The main findings of this study revealed that vulnerability severity scores very much depend on the accessor, even if an industry standard such as CVSS is being used or a very experienced accessor is conducting the scoring [12]. It implies, that using CVSS methodology is not enough, and that the accessor and the people responsible for vulnerability management should not fully rely on the scores depicted by systems and prioritize vulnerabilities also according to the assets and the corporate value that is at stake. This also aligns with the previously mentioned statement by the developers of CVSS, stating that using CVSS alone is not enough for a successful and secure vulnerability management.

A recent paper summarized some potential improvements for CVSS that were proposed in other scientific literature. One method excluded subjective factors of CVSS, however, failed to include the importance of the asset. Another method included a distribution model for evaluating the complexity of an exploitation and the availability of an appropriate way to patch the according vulnerability. However, this methodology failed to consider the affected asset as well. Also, the other two examples that were mentioned did not manage to improve the CVSS properly [13]. This indicates that even small improvements of the CVSS are not easily made.

Summarizing, it can be noted that the CVSS is a widely accepted approach and even considered an industry standard despite having its flaws and being criticized. However, it should be emphasized that the developers of CVSS are constantly evolving and improving this system in order to fulfil the expectations of its users and the responsibilities they indirectly might carry. Furthermore, it should be emphasized that the responsibility also lies within the correct behaviour of the user and within vulnerability managers making the final and appropriate prioritization decisions.

The following section will introduce alternative systems to the CVSS such as the Tripwire score that was created by a specialised company, and vulnerability systems based on open-source communities such as OWASP. The later discussion will compare these different methods and reflect on the scope and responsibilities of such scoring systems.

5.2 Alternative systems

Despite the fact that CVSS scoring is the predominant approach and even considered an industry standard, several organizations and companies have come up with their own methodology for the severity scoring of vulnerabilities. This section will primarily reflect on the solutions offered by Tripwire, a specialized company in the field of vulnerability management, and OWASP, an open-source community for cybersecurity related issues. The choice for these solutions is based on its popularity within the sector and the fact, that it has been mentioned in scientific literature as well [12].

The first alternative the section aims to reflect on is the vulnerability scoring system by Tripwire within their vulnerability management solution known as Tripwire IP360. Tripwire is a specialized company within the field of cybersecurity that offers, apart from their vulnerability management product line, cybersecurity related solutions such as enterprise security foundations and cloud-based infrastructures. It has to be noted that their vulnerability scoring system is based on the CVSS, however, includes additional user specific metrics in order to make the prioritization of vulnerabilities more tailored to the customer's environment. They note that especially large enterprises might suffer from many vulnerabilities with a critical CVSS score what makes the prioritization, despite using an industry standard, incredibly difficult [14]. The Tripwire Vulnerability Scoring System should, hence, not be seen as an alternative, but rather a separate product that makes use of CVSS and adds further information and advanced tools to this methodology. The determination of the risk is clustered into three main categories, so-called parameters being "Risk Class", "Skill Level", and "Vulnerability Age". Risk Class reflects on the potential consequences an exploitation of the vulnerability would have, the user involvement that is required, and, finally, the importance of the target application. Skill Level refers to the difficulty of a potential exploitation based on the availability of potential exploit methods. Lastly, Vulnerability Age refers to old, but well-known vulnerabilities as those are at greater risk to being exploited by automated malware tools. The Tripwire Vulnerability Scoring System is a complement to the CVSS, it is possible that even though all vulnerabilities received the same (critical) score based on the CVSS methodology, the severity of them varies according to the Tripwire Vulnerability Scoring System [15].

Overall, it can be noted that Tripwire provides additional accuracy to the CVSS within their own vulnerability scoring system. It further provides additional value and protection as well as it might result in a more successful vulnerability management. However, these features and tools are not free of charge, indicating that the more companies are willing to pay for their security, the greater the protection they receive. This relates to the points mentioned in the previous section about the context of vulnerability management in large enterprises and the unclear knowledge about cybersecurity budgeting decisions. In order to explore open-source versions of similar tools, the following paragraphs will analyse the solutions offered by the OWASP community.

While there is no explicit information available for a vulnerability scoring system made by OWASP, they, however, came up with an overall Risk Rating Methodology. This approach is split into six steps, with the fourth step labelled "Determining Severity of the Risk". Although it is important to view this rating methodology as a whole model that builds upon each step, this paper will precisely look at the particular step that refers to severity scoring. During this step, the user is required to combine the estimates of the likelihood of exploitation and the ones of the potential impact [16].

Important is the word "estimates", emphasizing that the risk rating methodology is rather a guide or methodology instead of an automized tool. Considering that this approach is followed by using manual estimates, it is very likely that the accuracy might be lower than the one provided within the Tripwire Scoring System. Furthermore, it does not provide an automized tool which could be necessary for large enterprises encountering lots of vulnerabilities on a frequent basis. However, an essential key aspect is its free usage. Similar to CVSS, the OWASP Risk Rating Methodology is free of charge, and definitely a better solution than not providing any risk methodology at all.

Another important benefit of using the OWASP methodology, compared to the one proposed by FIRST, is its simplicity, considering that there are less metrics/parameters involved for estimating the severity. Furthermore, it takes business factors more into consideration than technical ones, proposing a methodology that is more tailored to the enterprise that uses it. It also includes factors that are not included in the CVSS base score at all [13]. Considering these factors, OWASP might provide a good alternative to the most commonly used CVSS.

6 ORGANIZATIONAL FACTORS AND GOOD PRACTICES

While optimizing the technical operations of vulnerability management is an important task, managers should not underestimate the influence of human and organizational factors. To analyze these factors, common "pathways" to vulnerabilities [3] as well as patch management as a practice was included. These aspects were then reflected within the broader context of ISMS (Information Security Management Systems) which can be viewed as a process or cycle of the cybersecurity program of an organization [17].

There is a common saying that within the field of cybersecurity, the greatest vulnerability are humans. Although it has to be acknowledged that these vulnerabilities are most often interlinked with technological flaws as well, the key point of this statement might be true. These vulnerabilities can be things such as weak password choices, but also errors within organizational policies, incorrect managerial decision making, or a lack of cybersecurity related awareness within the organization. In fact, studies revealed that these aspects might even be correlated with each other. For instance, secure password choices are often correlated with a sufficient amount of awareness training. Another research also suggested that high workload and difficult tasks could negatively affect system states and performance as the appearance of human-errors becomes more likely [3].

Apart from general pathways to vulnerabilities, vulnerability management can also be addressed on a process level and within the context of Information Security Management Systems (ISMS). In this context, the aforementioned policy aspect carries an essential role and is not only a potential pathway to vulnerabilities, but also overall an important element of vulnerability management. Policies define the responsibilities and accountabilities and in order to retain their role as a strategic element, they not only need to be constantly evaluated and adjusted, but also communicated clearly. Further, tactical elements of vulnerability management include guidelines, processes, communication, the development of a plan, and patch management [17]. A document that was published by the National Institute of Standards and Technology (NIST) contains guidelines regarding patch management that enterprises should follow. It states that timing plays an immense role due to the limited resources of enterprises. Especially when it comes to the issue of prioritization, this aspect is of significant relevance. Considering that each patch should be tested before being deployed, the conflict between these three factors is not easily solved and, therefore, should be carefully considered when planning and executing the organization's vulnerability management program. While the document also refers to vulnerability scoring, it has to be noted that this is only a fraction of the whole matter of vulnerability prioritization and management [18]. Another paper that approached patch management by conducting a literature synthesis, also concluded that in terms of prioritization, enterprises should make decisions not only based on the according scores, but also on the context as this would result in a more accurate depiction of the overall risk [19].

Based on these arguments, it would be wrong to put the blame on organizations such as FIRST for not being able to develop the perfect severity scoring methodology that does all the work in an automized way. This would be a rather utopian way of approaching this matter. Good practices for vulnerability management involve more than technological aspects. Policies, awareness, and the correct measurements taken by non-cybersecurity managers are most likely as important. Furthermore, it should not be seen as separate tasks, but rather as a co-existence between technological methodologies, budgeting decisions, and organizational aspects.

7 DISCUSSION

The following discussion will reflect on both the CVSS and the alternative methods stated such as the Tripwire Vulnerability Scoring System and the OWASP Risk Rating Methodology by also reflecting on the role of the right budgeting decisions and organizational factors.

What can be drawn from the analysis is that the CVSS is the most-widely used methodology for calculating the severity of vulnerabilities. Furthermore, it is free to use and has a pioneering role for other vulnerability scoring systems such as the one by Tripwire. While it is, to some extent, criticized for not including relevant, organizational factors, the organization FIRST clearly emphasized within their user guide that the CVSS should not be used on its own and is not enough for a secure vulnerability management. Furthermore, the CVSS is steadily improving and updated with each new version that is introduced. However, there remains the question whether it is the CVSS' responsibility, as an industry standard, to meet the expectations of its users, considering that tools such as the Tripwire Vulnerability Scoring System seem to be able to provide more accurate results and that the need for accurate vulnerability management, especially in large enterprises, is significantly increasing.

Another point, that builds upon what was written in Section 3, deals with the question about the right budget. Vulnerability management is only a very small fraction of a functioning cybersecurity foundation and enterprises could be overwhelmed by all the optimizing options that exist. For instance, enterprises could manage their vulnerabilities using CVSS only, or by estimating it according to the risk rating methodology by OWASP. However, this would require experts using these methodologies and accordingly make the right decisions using this input. Also, if there is a large number of vulnerabilities, manual estimations and the according prioritization is burdensome. Alternatively, they could invest in solutions such as the one by Tripwire where they profit by more accuracy and automized processes. While Tripwire would most likely bring the most efficient results, enterprises need to make an appropriate decision: Investing enough in their vulnerability management, being in automized software or the right people, while simultaneously remaining profitable. Similar to other risk-related topics, there is this trade-off between the losses that could potentially arise due to these vulnerabilities being successfully exploited, and the expenses that arise when trying to mitigate the risk as much as possible.

While these technological and financial factors need to be considered, there are many other, organizational and human factors that influence the exploitation of vulnerabilities and potential attacks which should not be underestimated. The findings regarding the correlation between training, policies, and vulnerability exploitation, hence, suggest that vulnerability management is a more complex process than the optimization of severity scoring based on a mathematical formula. The responsibility lies not only within cybersecurity professionals developing appropriate and accurate systems, but also within managers and employees, knowing the value of their assets and the potential consequences of their vulnerabilities. The findings further revealed that even the subject of patch management, which is only a fraction of vulnerability management, is a rather complex issue as well and involves more factors such as timing, prioritization, and testing.

8 LIMITATIONS

The first limitation of this paper is its imbalance between the reflection of vulnerability severity scoring systems and the one about organizational and financial factors. During the research, it seemed that the CVSS and its alternatives have a predominant role and are frequently appearing in research papers when it comes to best practices regarding vulnerability management. Also, a lack of direct comparisons between these systems could be discovered, making a judgement based on scientific literature regarding the "best" system increasingly difficult. The main objectives such as reflecting on the CVSS could be fulfilled, however, the paper failed to provide a complete summary of most methods existing. This might be due to the fact that the field of vulnerability management involves too many factors that cannot be captured within a paper of this size without taking away necessary, detailed information.

9 IMPLICATIONS FOR FUTURE RESEARCH

The paper suggests that there is a lack of a detailed comparison and exploration of existing vulnerability scoring systems, supporting especially large enterprises at choosing an appropriate mean for handling their vulnerability management. It implies, that organizational and budget-related decisions have equal importance, and that the latter aspect seems to be a particularly unclear task for most companies. The question regarding best practices of vulnerability management should be covered as a whole, providing enterprises with essential information about severity scoring methodologies, asset prioritization, patch management practices, a sufficient cybersecurity budget, and the importance of training and policies.

10 CONCLUSION

Overall, there is no doubt that vulnerability management has an important role within the cybersecurity program of large enterprises. What could be discovered during the research, is that companies might struggle to quantify and prioritize their vulnerabilities properly due to the fact that there are many factors that need to be taken into consideration. While methodologies such as the CVSS and the OWASP Risk Rating Methodology remain reliable, free-of-charge solutions, risk-minimizing goals require enterprises to invest into the right people, processes and, ideally, automized software to support them at quantifying their vulnerabilities' severity. Although risk scoring is an important part in vulnerability management, it has to be noted that these methodologies might not take crucial factors into consideration. Despite the fact that these solutions are constantly improving and becoming more and more accurate, organizational factors influence this matter a lot as well. Aspects such as policies, risk related awareness, and communication play a crucial role and come along with benefits that severity scorings alone sometimes fail to provide. Therefore, the findings of this paper suggest an alignment of all these factors, indicating that the application of vulnerability scorings is an important part, however, not sufficient if used on its own. More precisely, vulnerability scoring systems should be embedded within well-designed processes and patch management programs, by taking into account the respective assets.

REFERENCES

[1] Kekül, H., Ergen, B. & Arslan, H., A multiclass hybrid approach to estimating software vulnerability vectors and severity score. *Journal of Information Security and Applications*, **63**, 103028, 2021. DOI: 10.1016/j.jisa.2021.103028.

[2] Holm, H. & Afridi, K., An expert-based investigation of the Common Vulnerability Scoring System. *Computers and Security*, **53**, pp. 18–30, 2015. DOI: 10.1016/j.cose.2015.04.012.

[3] Kraemer, S., Carayon, P. & Clem, J., Human and organizational factors in computer and information security: Pathways to vulnerabilities. *Computers and Security*, **28**(7), pp. 509–520, 2009. DOI: 10.1016/j.cose.2009.04.006.

[4] Bharwani, S., Vulnerability definitions. 2020. https://www.weadapt.org/knowledge-base/vulnerability/vulnerabilitydefinitions. Accessed on: 20 Nov. 2021.

[5] Haimes, Y., On the definition of vulnerabilities in measuring risks to infrastructures. *Risk Analysis*, **26**(2), pp. 293–296, 2006. DOI: 10.1111/j.1539-6924.2006.00755.x.

[6] Statista, published by Accenture, Average annual costs for external consequences of cyber attacks on global companies in 2018, 2018. https://www.statista.com/statistics/241255/main-consequences-of-cyber-attacks-in-selected-countries. Accessed on: 20 Nov. 2021.

[7] Uddin, H., Hamid, A. & Hassan, M., Cybersecurity hazards and financial system vulnerability: A synthesis of literature. *Risk Management*, **22**(4), pp. 239–309, 2020. DOI: 10.1057/s41283-020-00063-2.

[8] FIRST, Mission Statement, 2021. https://www.first.org/about/mission. Accessed on: 20 Jan. 2022.

[9] FIRST, Common Vulnerability Scoring System v.3.1: Specification Document – Revision 1, 2021. https://www.first.org/cvss/v3.1/specification-document. Accessed on: 17 Jan. 2022.

[10] FIRST, Common Vulnerability Scoring System: User Guide, 2021. https://www.first.org/cvss/user-guide. Accessed on 17. January 2022.

[11] Allodi, L., Biagioni, S., Crispo, B., Labunets, K., Massacci, F. & Santos, W., Estimating the assessment difficulty of CVSS environmental metrics: An experiment. *Future Data and Security Engineering*, **10646**, Springer: Cham, pp. 23–39, 2017. DOI: 10.1007/978-3-319-70004-5_2.

[12] Allodi, L., Cremonini, M., Massacci, F. & Shim, W., Measuring the accuracy of software vulnerability assessments: Experiments with students and professionals. *Empirical Software Engineering*, **25**(2), pp. 1063–1094, 2020. DOI: 10.1007/s10664-019-09797-4.

[13] Pecl, D., Safonov, Y., Martinasek, Z., Kacic, M., Almer, L. & Malina, L., Manager asks: Which vulnerability must be eliminated first? *Lecture Notes in Computer Science (including subseries Lecture Notes in Artificial Intelligence and Lecture Notes in Bioinformatics)*, **12596**, pp. 146–164, 2021. DOI: 10.1007/978-3-030-69255-1_10.

[14] Khimji, I. published via Tripwire, Tripwire vulnerability risk metrics: Connecting security to the business, 2019. https://www.tripwire.com/-/media/tripwiredotcom/files/white-paper/tripwire_vulnerability_risk_metrics_white_paper.pdf?rev=7d6ee34728c24342811273edc cc69619. Accessed on: 20 Jan. 2022.

[15] Tripwire, Advanced vulnerability risk scoring and prioritization, 2019. https://www.tripwire.com/-/media/tripwiredotcom/files/solution-brief/tripwire_advanced_vulnerability_risk_scoring_and_prioritization_solution_brief.pdf?rev=85736590681147dfbc7423df7b9b5436. Accessed on: 20 Jan. 2022.

[16] OWASP, OWASP risk rating methodology, 2021. https://owasp.org/www-community/OWASP_Risk_Rating_Methodology. Accessed on: 20 Jan. 2022.
[17] Nyanchama, M., Enterprise vulnerability management and its role in information security management. *Inf. Secur. J. A Glob. Perspect.*, **14**(3), pp. 29–56, 2005.
[18] Souppaya, M. (NIST) & Scarfone, K., Guide to Enterprise Patch Management, 2013. https://nvlpubs.nist.gov/nistpubs/SpecialPublications/NIST.SP.800-40r3.pdf. Accessed on: 20 Jan. 2022.
[19] Dissanayake, N., Jayatilaka, A., Zahedi, M. & Babar, M., Software security patch management: A systematic literature review of challenges, approaches, tools and practices. *Information and Software Technology*, **144**, 106771, 2022. DOI: 10.1016/j.infsof.2021.106771.

SECTION 4
SAFETY AND SECURITY ENGINEERING

EMOTIONAL ANALYSIS OF SAFENESS AND RISK PERCEPTION OF DIFFERENT PAYMENT SYSTEMS IN ITALY AND THE UK DURING THE COVID-19 PANDEMIC

FABIO GARZIA[1,2,3], FRANCESCO BORGHINI[1], JESSICA BOVE[1],
MARA LOMBARDI[1] & SOODAMANI RAMALINGAM[4]
[1]Safety & Security Engineering Group – DICMA, SAPIENZA – University of Rome, Italy
[2]Wessex Institute of Technology, Southampton, UK
[3]European Academy of Sciences and Arts, Austria
[4]School of Physics, Engineering and Computer Sciences, University of Hertfordshire, UK

ABSTRACT

Public safety, security and risk perception is an important aspect considered in opinion mining and sentiment analysis typically carried out on social networks. This involves considering each individual's opinion and determining a sense of what the public feels about an incident, event or place. In that sense, social networks play an important role in capturing the emotions of people. Security and safety managers can employ opinion mining and sentiment analysis as a tool to discover any unforeseen vulnerabilities in a precise manner and thereby plan and manage any associated risks. Furthermore, a continuous evaluation of risk perception can be carried out for timely and planned interventions in a seamless, effective manner to reduce or avoid any panic amongst communities. Without such advance techniques, safety and security of people, infrastructures and specific contexts can be easily compromised. Recent work in this direction has shown promising results in managing risks, especially during the COVID-19 pandemic. The purpose of the present work is to investigate the perception of risk associated with different payment systems, in Italy and the UK, during the COVID-19 pandemic, from 10 November 2020 to 13 May 2021, by means of the semantic analysis of the textual contents existing in Twitter.

Keywords: perceived risk assessment for security, open-source intelligent techniques for security, OSINT, opinion mining for security, sentiment analysis for security, payment systems risk perception.

1 INTRODUCTION

Public safety, security and risk perception is an important aspect considered in opinion mining and sentiment analysis typically carried out on social networks. This involves considering each individual's opinion and determining a sense of what the public feels about an incident, event or place. In that sense, social networks play an important role in capturing the emotions of people.

Security and safety managers can employ opinion mining and sentiment analysis [1]–[8] as a tool to discover any unforeseen vulnerabilities in a precise manner and thereby plan and manage any associated risks. Further, a continuous evaluation of risk perception can be carried out for timely and planned interventions in a seamless, effective manner to reduce or avoid any panic amongst communities. Without such advance techniques, safety and security of people, infrastructures and specific contexts can be easily compromised. Recent works in this direction has shown promising results in managing risks and especially during the COVID-19 pandemic [9]–[11].

The purpose of the present work is to investigate the perception of risk associated with different payment systems, in Italy and the UK, during the COVID-19 pandemic, from 10 November 2020 to 13 May 2021, by means of the semantic analysis of the textual contents existing in Twitter.

WIT Transactions on The Built Environment, Vol 214, © 2022 WIT Press
www.witpress.com, ISSN 1743-3509 (on-line)
doi:10.2495/SSR220111

The set of opinions to be employed for this purpose involves seeking across several open sources (Open-Source INTelligence (OSINT)) and therefore the processing of huge volumes of data (Big Data) in digital form from which to obtain information and knowledge. In the present work, Twitter was employed as a source and a proper estimation of the keywords included in the different tweets was done.

The keywords have been selected taking into consideration that the estimation of perceived risk is greatly correlated to the psychological characteristics, through the emotional reactions aroused by a certain physical and logical context.

2 PAYMENTS SYSTEMS

Payment systems are instruments and procedures aimed at reproducing the material movements of money from one subject to another, in order to regulate established economic transactions [12]. Through payment systems, goods and services can be purchased, disbursing a sum of money on the basis of agreements made and delivering this amount to a creditor counterpart. The payment systems sector has become an important part of the exchange activities of each country: with the passage of time it has become increasingly complex, requiring an ever greater choice between tools and technological solutions, suitable for promoting speed and efficiency of the exchange procedures.

Cash (banknotes and coins) allows immediate transfer between two subjects and is immediately reusable. Cash is generally used to carry out transactions of limited amount between physically present subjects; it generally guarantees the anonymity of transactions. The "legal tender" of banknotes and coins involves the acceptance of cash, without prejudice to any limits set by law to regulate any type of transaction.

Monetary exchanges now take place largely through technological systems. The subjects that manage them are increasingly involved in the control of data networks and information technology (IT) systems [13]. The main types of electronic payment used are represented by: credit cards, debit cards, bank transfer, direct debits, money transfers, prepaid cards, digital payments (mobile payments, E-wallet payments, etc.), contactless payments.

2.1 Payments in Italy

Regarding Italy, based on the data collected by the European Central Bank (ECB) up to 2016, it was found that banknotes and coins are used by Italians in 85.9% of cases and cards are used only in 12.9% of transactions. If we consider the economic value of the transactions, the cards move 28.6% of the total, while the cash 68.4%, it can be deduced that cash is used for payments of a small amount on average.

Starting during the pandemic lockdown with an increase in e-commerce (eight out of ten Italians made use of it), the intensification of the use of credit cards continued in the post lockdown and is destined to continue. Today, the average frequency of electronic payments is around 2.9 times a month, compared to 2.6 in 2019 and 2.2 in 2018.

During the period corresponding to the medical lockdown, the use of electronic money by Italians increased with a jump comparable to that recorded in the eight years from 2011 to 2019. In the days of maximum emergency, the percentage of the value of the expenditure paid with credit, debit or loyalty cards, it went from 57%, an average figure recorded in 2019, to 68% [14].

2.2 Payments in the UK

Until a few years ago, in the UK it was still cash that moved the economy: in 2007, banknotes accounted for 60% of transactions, more than half. Ten years later there was overtaking: for the first time, debit cards exceeded cash. In 2019, the share of cash transactions carried out fell to 30% [15].

The trend of favouring cashless payments has been accelerated by the pandemic state, predicting that in the short term the percentage of the UK population who will still rely on cash will be 20%.

3 SEMANTIC ANALYSIS METHOD

Semantic analysis is somewhat diverse from the topic-based text classification where classification is carried out by means of predefined topics and the topic related words are crucial. Instead, opinion words such as positive, negative, or neutral are given importance in semantic analysis. In this work, we take into consideration a sentence-level classification as expressing a positive or negative opinion. We examine key tools followed by the method employed in this work:

1. Twitter: Twitter has been selected as the data source as it is one of the most popular microblogging and social networking site used by people today. Twitter revolves around the principle of followers. This tends to influence the opinion of others. From a Machine Learning perspective, Twitter messages are convenient to use as they are restricted to 280 characters/tweet and lends itself to a simple, structured way of analysing the sentence effectively. In this work, tweets in English and Italian were considered that relate to the considered context. Our aim is to assess the emotions of people and estimate their risk perception of different payment systems during COVID-19 pandemic.
2. OSINT tool: We consider the open-source intelligence (OSINT) software which relates to gathering data from open sources to produce actionable intelligence that is typically used by analysists using non-sensitive intelligence. As we use Twitter as our online source, we choose a set of keywords from Tweets that relate to perceived risk by people. These keywords correlate to an emotional experience of an individual for a specific context. The work is tested on Twitter as illustrated in the following, during COVID-19 pandemic, from 10 November 2020 to 13 May 2021.
3. Python, Twython, NLTK, Text2emotion [16]: Python is a high-level, interpreted, general-purpose programming language. Its design philosophy highlights code readability with the usage of considerable indentation. Python is dynamically-typed and garbage-collected. It supports multiple programming paradigms, together with structured (particularly procedural), object-oriented and functional programming. It is often defined as a "batteries included" language because of its comprehensive standard library. Twython is a Python library providing an easy way to access Twitter data, also allowing to determine the research area (Italy and the UK in our case). Natural Language Toolkit (NLTK) is a Python package for processing and analysing text, from basic functions to sentiment analysis powered by machine learning. Text2emotion is the python package to extract the emotions from the content, processing any textual message and recognize the emotions embedded in it, classifying them between the five emotions "Joy", "Anger", "Surprise", "Fear", "Sadness". These programming language/packages have been used to perform a part of the analysis which is compared with results obtained from a survey for word association illustrated in the following, where participants are asked to associate every selected words before to the hedonic valence (positive, negative, neutral) and then to one of the eight emotions "Joy", "Anger", "Surprise", "Fear",

"Sadness", "Anticipation", "Disgust", "Attraction". The keywords utilized as seek word are shown in Table 1. Table 2 provides typical categories of the one thousand more repeated words related to colleting period and the frequencies of occurrences of these words.

4. Procedure: First, a group of positive and negative words are defined. In addition, a group called "Neutral words" is included. It considers the words which do not have positive or negative valence. Next, a survey for word association with emotion is conducted. For the two languages considered, a group classification of the script sentences is carried out. For this purpose, a suitable survey was designed, and inputs collected from 40 evenly distributed male and female participants, aged between 17 and 75 for Italy and the same for the UK. This survey was later analysed which gave an insight into how people map words with hedonic valence (positive, negative, neutral and with the eight primary emotions before and with only five primary emotions, indicating the remaining ones as "others" to make these results comparable with results obtained by means of Python software packages. Such a mapping was obtained for every word considered in this work. They are also compared with results obtained by means of NLTK, Text2emotion.

Table 1: Summary of the English and Italian keywords used as search terms.

Country	Primary keyword	Secondary keywords 1	Secondary keywords 2
Italy	Italy-covid	money, credit cards, credit card, bancomat, coins, banknotes, cash, debit cards, debit card	virus, coronavirus, pandemic, covid, covid-19, covid 19, sars-cov-2
Italy	Italy	money, credit cards, credit card, bancomat, coins, banknotes, cash, bank, banks, post offices, current account, debit card, debit card	–
UK	UK-covid	money, credit card, atm card, coins, cash, banknotes, debit card	virus, coronavirus, covid, covid-19, covid 19, sars-cov-2
UK	UK	money, credit card, atm card, coins, cash, banknotes, bank, banks, bank account, debit card	–

Table 2: Total number of words collected over the one thousand more recurrent words.

Word	Count (English)	Count (Italian)
Positive words	66,690	16,160
Negative words	26,779	11,960
Neutral words	602,873	221,254
Total words	696,342	249,374

4 RESULTS OF SENTIMENT AND EMOTIONAL ANALYSIS FROM TWITTER

The survey started during the COVID-19 pandemic, from 10 November 2020 to 13 May 2021. The managing of the data involved generated a huge quantity of information associated to the studied collecting period (1,162,157 in Italy and 2,203,676 words in the UK, for a total of 3,365,833 words). The total number of words collected over the one thousand more recurrent words are illustrated in Table 2.

The results of hedonic valence evaluated by means of NLTK are shown in Fig. 1, considering only the secondary keywords of "cards" and "cash". As it is possible to see from Fig. 1, with the exception of words collected with primary keywords "Italy-covid", there is a general positive hedonic valence for all the words collected with the other primary and secondary keywords.

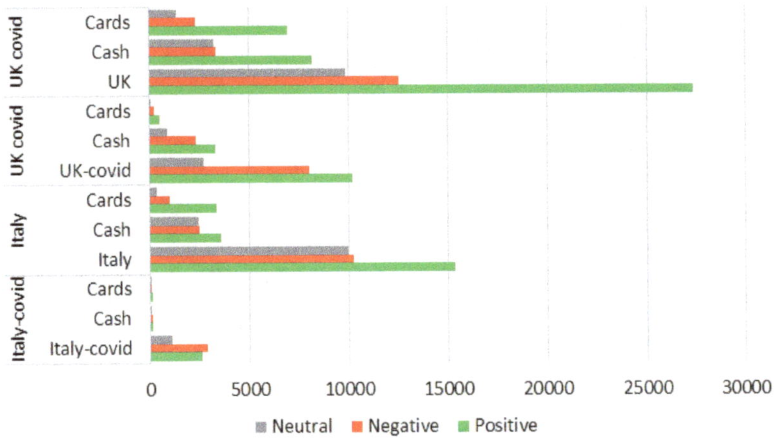

Figure 1: Hedonic valence for primary keywords considering only the secondary keywords of "cards" and "cash" for Italy and the UK.

In order not to overburden the questionnaire to be administered to the sample of people considered, only the eight words regarding cash and eight words regarding cards, for Italy and the UK respectively, positive and/or negative, were considered, both for the English language and for the Italian language. The considered words, in English and Italian, are shown in Table 3.

Table 3: Considered English and Italian words considered for the survey.

Word regarding "cash" in English	Word regarding "payment cards" in English	Word regarding "cash" in Italian	Word regarding "payment cards" in Italian
pandemic	pay	euro	alone
people	use	alone	euro
like	payment	pay	cashback
new	like	without	pay
need	new	home	debt
back	back	do	do
time	details	way	payment
UK	need	Italy	way

Fig. 2 illustrates the distribution of the number of tweets across the studied timeline of the four primary keywords (Italy-covid, Italy, UK-covid, UK) during the data collecting period.

We can see a descending/oscillating trend regarding the curves related to "Italy-covid" and "UK-covid", due to reduced restrictions in view of next summer season. We can also see a oscillating trend around a mean value regarding the curves related to "Italy" and "UK", as these curves are not directly related "covid" as primary keyword.

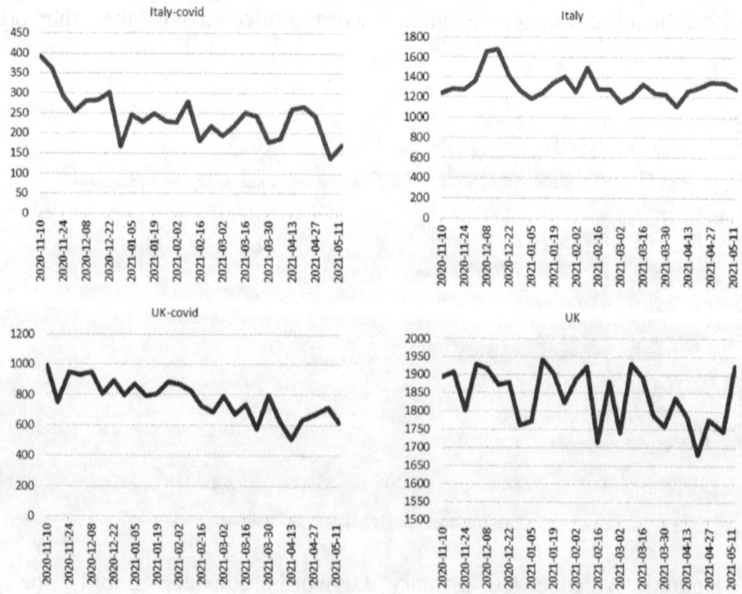

Figure 2: Distribution of the number of tweets across the studied timeline of the four primary keywords (Italy-covid, Italy, UK-covid, UK) during the data collecting period.

Each of the four survey results related to "cash" and "cards", for Italy and the UK, respectively, are illustrated in the following, together with a comparison with results obtained by software analysis packages.

4.1 Survey results for "cash" in Italy

Regarding the survey results for "cash" in Italy and the related words used for the survey, we consider the hedonic valence and the distribution of each primary emotion (for eight and five+ others emotions survey) as shown in Figs 3 and 4, respectively. From Fig. 3 we can see that there is a general positive valence for all the survey words, except for "alone", "pay", and "without", as expectable. From Fig. 4, we can see that the predominant positive emotion is "Joy" for both surveys. Negative emotions represent a minority with respect to other positive emotions.

4.2 Survey results for "cards" in Italy

Regarding the survey results for "cards" in Italy and the related words used for the survey, we consider the hedonic valence and the distribution of each primary emotion (for eight and five+ others emotions survey) as shown in Figs 5 and 6, respectively. From Fig. 5 we can see

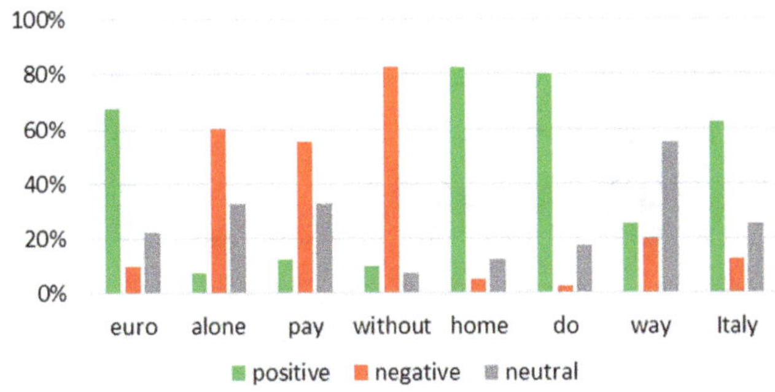

Figure 3: Hedonic valence for the considered survey words regarding "cash" in Italy.

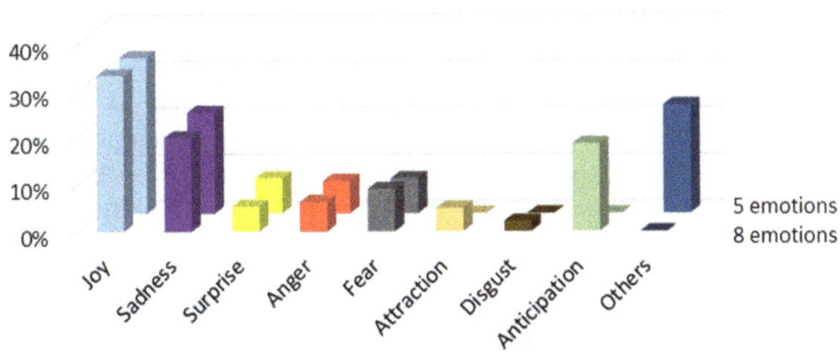

Figure 4: Distribution of each primary emotion (for eight and five+ others emotions survey) for the considered survey words regarding "cash" in Italy.

that there is a general positive valence for all the survey words, except for "alone", "pay", and "debt", as expectable. From Fig. 6, we can see that the predominant positive emotion is "Joy" for both surveys. Negative emotions represent a minority with respect to other positive emotions.

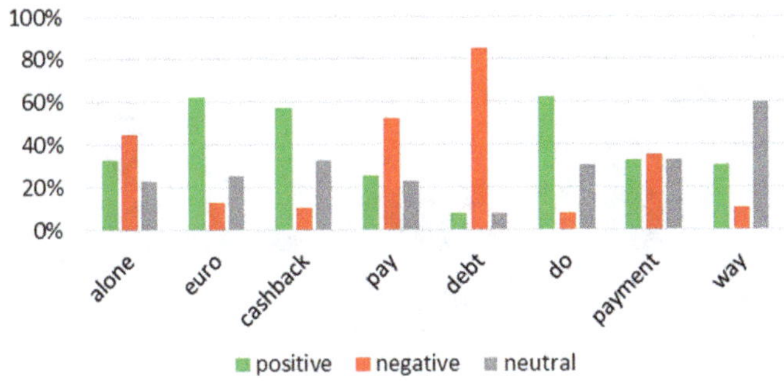

Figure 5: Hedonic valence for the considered survey words regarding "cards" in Italy.

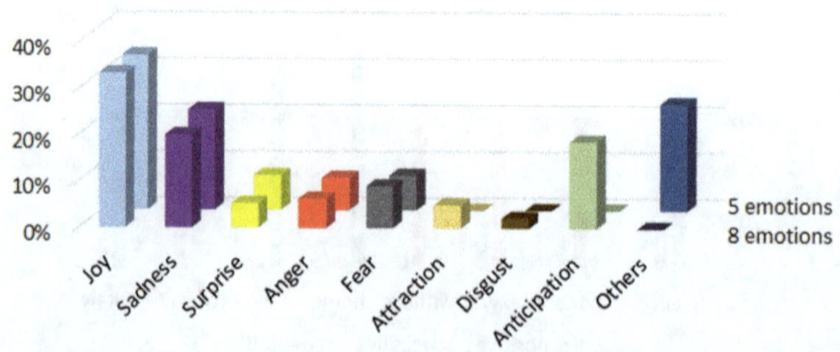

Figure 6: Distribution of each primary emotion (for eight and five+ others emotions survey) for the considered survey words regarding "cards" in Italy.

4.3 Survey results for "cash" in the UK

Regarding the survey results for "cash" in the UK and the related words used for the survey, we consider the hedonic valence and the distribution of each primary emotion (for eight and five+ others emotions survey) as shown in Figs 7 and 8, respectively. From Fig. 7 we can see that there is a general positive valence for all the survey words, except for

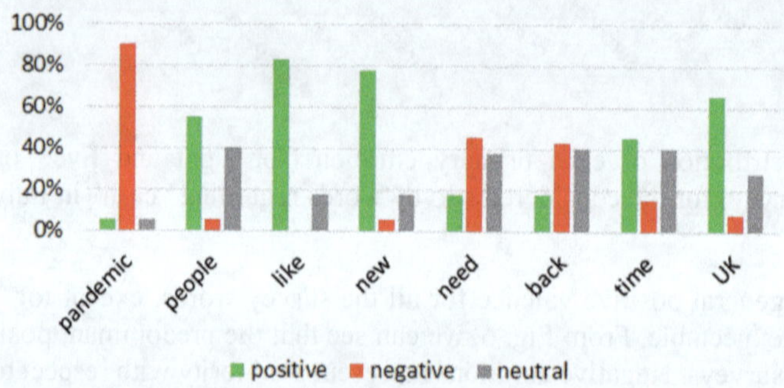

Figure 7: Hedonic valence for the considered survey words regarding "cash" in the UK.

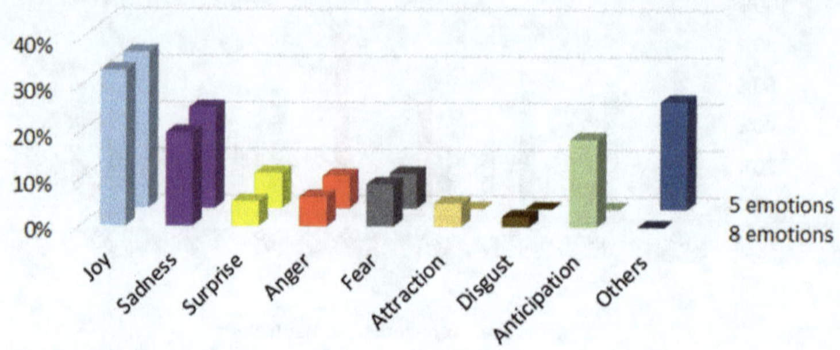

Figure 8: Distribution of each primary emotion (for eight and five+ others emotions survey) for the considered survey words regarding "cash" in the UK.

"pandemic", "need", and "back", as expectable. From Fig. 8, we can see that the predominant positive emotion is "Joy" for both surveys. Negative emotions represent a minority with respect to other positive emotions.

4.4 Survey results for "cards" in the UK

Regarding the survey results for "cards" in the UK and the related words used for the survey, we consider the hedonic valence and the distribution of each primary emotion (for eight and five+ others emotions survey) as shown in Figs 9 and 10, respectively. From Fig. 9 we can see that there is a general positive valence for all the survey words, except for "payment", "back" and "need", as expectable. From Fig. 10, we can see that the predominant positive emotion is "Joy" for both surveys. Negative emotions represent a minority with respect to other positive emotions.

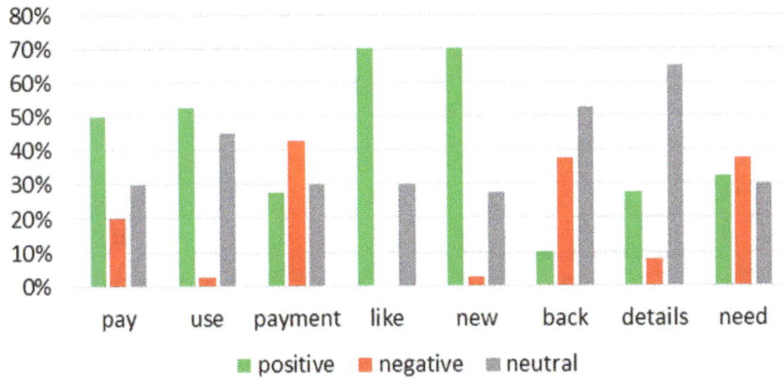

Figure 9: Hedonic valence for the considered survey words regarding "cash" in the UK.

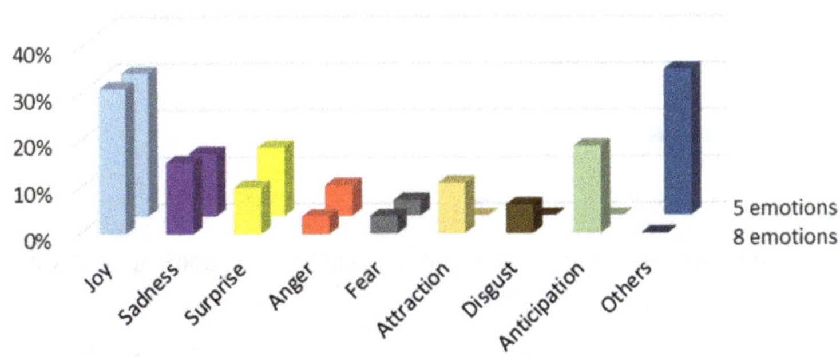

Figure 10: Distribution of each primary emotion (for eight and five+ others emotions survey) for the considered survey words regarding "cash" in the UK.

4.5 Survey results and comparison between survey and software analysis packages results

The comparison between survey and software analysis packages results are shown in Figs 11 and 12 for NLTK and Text2emotion respectively.

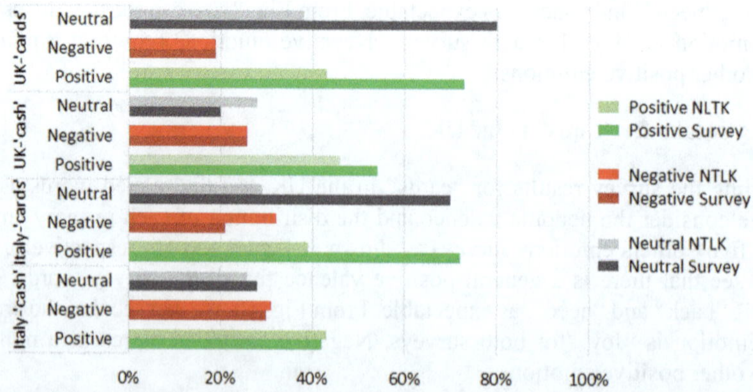

Figure 11: Comparison between survey and NLTK analysis software results.

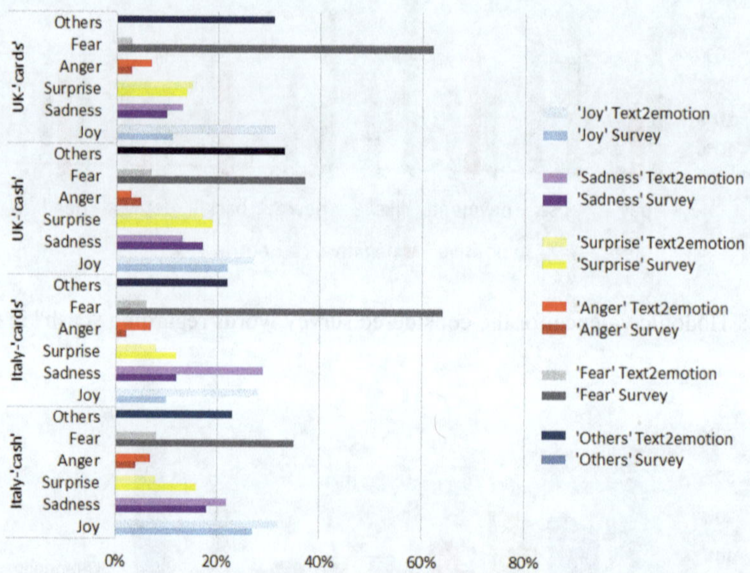

Figure 12: Comparison between survey and Text2emotion analysis software results.

Regarding only survey results shown in previous paragraphs, it is possible to state that:

1. For the analysis carried out on all the tweets collected in Italy and analysed through NLTK and Text2emotion, the hedonic value aroused both by the texts referring to cash and to cards and ATMs is positive and, although for both situations analysed the prevailing emotion is "Fear", for tweets referring to cash payment systems this is immediately followed by "Joy".
2. For the analysis carried out on all the tweets collected in the UK and analyzed through NLTK and Text2emotion, the hedonic value aroused both by the texts referring to cash payment systems and to those with cards and ATMs is positive and the prevailing emotion aroused is "Fear".

3. On the basis of the questionnaires administered in Italy, the prevailing sentiment for the most repeated words associated with cash is positive, arousing mainly “Joy”. For the most repeated words in the tweets referring to the use of cards and ATMs, there was a negative feeling, represented by “Sadness”. In both situations there was a continuity between the choice made with an available list of eight emotions and the one in which it was possible to choose only between five.
4. For the questionnaires administered in the UK, the hedonic value associated with the most repeated words, found both in the tweets referring to cash and cards, is positive. The emotions aroused turn out to be “Joy” and “Anticipation” in the first case and only “Joy” in the second. We want to highlight that there was no continuity in the responses detected when the available choice was only between five emotions instead of eight. It can be hypothesized that this is possible both for the ambivalent emotion aroused by some words, and because even the list of the eight primary emotions, despite being more complete, may not contain the nuances of emotions that can be experienced by the user in relation to a specific word, or, more likely, for the common use of complex emotions to replace the primary components that determine them.

Regarding the results of comparison between survey and software analysis packages shown in Figs 11 and 12, it is possible to state, for all four considered cases, that:

5. The trend of the hedonic valence shows a similar behavior both for the results obtained by the software packages and by the questionnaires administered, with a predisposition to positive sentiment.
6. a discrepancy can be observed between the results obtained through the analysis carried out using software, in which the predominant emotion was found to be “Fear”, and the results obtained from the questionnaires, where the most encountered emotion turns out to be “Joy”. It is important to highlight how often the most used choice in the proposed survey was “Others”, probably because, as already pointed out, it is more common to use complex emotions to replace the primary components that determine them.

The results show how the selected samples, respectively representative of the perception of users in Italy and the UK, present a tendency for the most part to a positive value, which is reflected in the propensity to “Joy”, except for the perception identified in the section corresponding to perception of the use of cards and ATMs by Italian users. This propensity could depend on resilience memory. In psychology, resilience is a word that indicates the ability to deal positively with traumatic events, adapting to individual situations and constantly changing. This phenomenon explains the optimism found in the texts analyzed, despite following a difficult and stressful situation such as that of the COVID-19 pandemic and the resulting changes and restrictions. A difference that can be found between the samples of Italy and the UK is that in the second case analyzed an adequate match was not found between some of the words and the emotions selected and proposed in the questionnaires, both in the eight and in the five emotions model. There are no other particular differences in the perception of the risk found for the use of the different payment systems in the two countries considered, probably because the course of the emergency situation, albeit with some different nuances, does not show obvious discrepancies.

These results confirm the existence of an underlying optimism that causes emotions not justified by the circumstances, to underline an inconsistency between the perception of reality and reality itself. In fact, although the studies in this regard have not been properly defined, the objective risk from the point of view of safety and security linked to the use of the various payment systems is not negligible and therefore one would expect, contrary to the results obtained, a negative hedonic value, associated with a negative emotion.

5 CONCLUSIONS

A methodology for the assessment of the perception of risk associated with different payment systems, in Italy and the UK, during COVID-19 pandemic, from 10 November 2020 to 13 May 2021, by means of the semantic analysis of the textual contents existing in Twitter has been illustrated.

It represents a valuable means when it is necessary to evaluate the perceived risk vs. the objective risk, in particular when we deal with critical contexts such as the considered one.

REFERENCES

[1] Medhat, W., Hassan, A. & Korashy, A.H., Sentiment analysis algorithms and applications: A survey. *Ain Shams Engineering Journal*, pp. 1093–1113, 2014.

[2] Ramalingam, S., Metadata extraction and classification of YouTube videos using sentiment analysis. *Proc. of IEEE International Carnahan Conference on Security Technology*, Orlando, FL, 2016.

[3] Cambria, E., Schuller, B., Xia, Y. & Havasi, C., New avenues in opinion mining and sentiment analysis. *IEEE Intelligent Systems,* **28**(2), pp. 15–21, 2013.

[4] Anastasia, S. & Budi, I., Twitter sentiment analysis of online transportation service providers. *Proc. of International Conference on Advanced Computer Science and Information Systems,* pp. 359–365, 2016.

[5] Shaikh, T. & Deshpande, D., A review on opinion mining and sentiment analysis. *International Journal of Computer Applications (0975-8887)*, 2016.

[6] Mayor, E. & Bietti, L.M., Twitter, time and emotions. *R. Soc. Open Sci.*, **8**, 201900, 2021.

[7] Liu, B., *Sentiment Analysis: Mining Opinions, Sentiments, and Emotions*, Cambridge University Press, 2015.

[8] Vijarania, M., *Sentiment Analysis: A Combined Approach*, Lambert Academic Publishing, 2020.

[9] Garzia, F., Borghini, F., Lombardi, M., Zita, M. & Ramalingam, S., Emotional analysis of safeness and risk perception of London and Rome airports during the COVID-19 pandemic. *Proc. of IEEE International Carnahan Conference on Security Technology*, pp. 1–6, 2021.

[10] Garzia, F., Borghini, F., Lombardi, M., Canfora, L. & Ramalingam, S., Emotional analysis of safeness and risk perception of London and Rome railway stations during the COVID-19 pandemic. *Proc. of IEEE International Carnahan Conference on Security Technology*, pp. 1–6, 2021.

[11] Garzia, F., Borghini, F., Lombardi, M., Moretti, A. & Ramalingam, S., Emotional analysis of safeness and risk perception of transports and travels by car and motorcycle in London and Rome during the COVID-19 pandemic. *Proc. of IEEE International Carnahan Conference on Security Technology*, pp. 1–6, 2021.

[12] Rambure, D. & Nacamuli, A., *Payments Systems,* Palgrave Macmillan, 2008.

[13] Bezovski, Z., The future of the mobile payment as electronic payment system. *European Journal of Business and Management*, **8**(8), pp. 127–132, 2016.

[14] European Central Bank, *Study on the Payment Attitudes of Consumers in the Euro Area (SPACE),* European Central Bank, 2020.

[15] UK Finance, *UK Payment Markets Summary 2021*, UK Finance, 2021.

[16] Python, Python Software Foundation, www.python.org/. Accessed on: 12 Aug. 2022.

PROMOTING SYSTEM SAFETY AND RELIABILITY THROUGH RISK QUANTIFICATION/VISUALISATION METHODOLOGY

TAKAFUMI NAKAMURA
Daito Bunka University, Japan

ABSTRACT
In this paper, a meta-methodology for holistically examining system failures is proposed to prevent their further occurrence. This methodology was introduced as a meta-methodology called system of system failures (SOSF). SOSF is represented in a three-dimensional space. In addition, a topological method used to monitor failure events within an SOSF space was presented to visualise the trajectory of system failures. A method was developed for quantifying the risk factors for a system failure that enables the factors to be quantified, monitored, and compared among the systems, and whose usage promotes system safety and reliability. The method was introduced using an interaction and coupling (IC) chart based on normal accident theory. An IC chart is used to classify object systems based on an interaction (linear or complex) and coupling (tight or loose); however, its effectiveness is limited by a subjective classification. The proposed method quantitatively (i.e. objectively) measures the risk factors and thus compensates for the subjectivity of an IC chart. Application examples in information and communication technology (ICT) engineering demonstrate that the proposed method applied to quantitatively monitor the risk factors helps improve the safety and quality of various object systems.
Keywords: risk management, system failure model, normal accident theory, interaction and coupling chart, information and communication technology.

1 INTRODUCTION

In this study, a new methodology is proposed to enhance the safety and reliability of a system by visualising its risk over time. The methodology used for understanding the current state of a failure has numerous shortcomings [1]. The first is the lack of a common language for discussing a failure, making it difficult for stakeholders involved in a failure to have a common understanding of the problem. Consequently, similar failures can occur repeatedly. A meta-methodology called system of system failures (SOSF) was introduced to address this first issue. Second, it is difficult to understand how the nature of a failure changes over time, rather than understanding each failure separately. For the second issue, SOSF is recognised as a failure space, and a topology is introduced into the space allowing individual failures to be recognised with relevance rather than an individual understanding. In this study, the first and second shortcomings were interlinked and organised to improve the safety and reliability of the system. Finally, the new methodology was applied to ICT failures, and its effectiveness was confirmed. ICT was chosen because its environmental changes are severe. These changes were identified in Gartner's study on IT trends [2]. There are three major concerns related to ICT risk: virtualisation, fabric technology, and big data. These three concerns will remain the same for 2022. An applied example shows that the risk quantification/visualisation methodology helps improve the safety and reliability of ICT systems by capturing the evolution and increasing the complexity of ICT technology with virtualisation technology. This application opens up the possibility of an expansion to other social systems.

This study is composed of five sections. Section 2 describes a survey of current methodologies. Section 3 introduces the new SOSF meta methodology, which overcomes the shortcomings of current approaches. Section 4 introduces new metrics into the SOSF space for a topological representation of system failure risk factors. Section 5 provides an

WIT Transactions on The Built Environment, Vol 214, © 2022 WIT Press
www.witpress.com, ISSN 1743-3509 (on-line)
doi:10.2495/SSR220121

application of ICT system failures. Finally, Section 6, concludes this paper by describing the effectiveness of this methodology and areas of future research.

2 SURVEY OF CURRENT METHODOLOGIES

2.1 Features of existing structuring methodologies and risk analysis techniques

Two methodologies are widely used, i.e. failure mode effect analysis (FMEA) [3] and fault tree analysis (FTA) [4]. FMEA reveals in a table form the linear relationship between the causes and consequences that lead to the final event in a bottom-up manner. By contrast, FTA is a method for clarifying the causes of a final event in a top-down manner as a logic diagram.

Both methodologies are primarily employed in the design phase. However, they are heavily dependent on personal experience and knowledge. FTA, in particular, tends to miss some failure modes among failure mode combinations, particularly emergent failures. There are two main reasons for missing failure modes. First, FMEA and FTA are rarely applied simultaneously. Accordingly, the sufficiency of the identified elements is not ensured in a mutually exclusive or collectively exhaustive manner. As the second reason, current approaches use a linear link between cause and effect, making it difficult to see complex issues involving many different stakeholders. Other major risk analysis techniques (including FMEA and FTA) have been described in various studies [5]–[7]. Most failure analyses and studies are based on either FMEA or FTA.

The existing structuring methodologies and risk analysis techniques do not sufficiently utilise a holistic approach. Many current methodologies connect cause and effect in a linear manner, and cannot respond appropriately to problems surrounded by many different stakeholders or problems under severe external changes in the environment. Therefore, what is lacking in the current majority of methodologies is the capability to tackle emergent problems caused by the complex relationships between the many stakeholders as well as external environmental changes surrounding the problem. A typical methodology applying a holistic approach is a soft systems methodology (SSM) [8]. SSM can manage emergent properties and thus implement preventative measures. Unlike the other current methodologies, SSM solves the above problem by revealing the nature of the stakeholders surrounding the problem among the customers, actors, transformers, owners, and worldview. Current methodologies tend to lose their holistic view of the root causes of a system failure. In addition, although most of them may be able to clarify the problem structures to confirm the effectiveness of a preventative measure, they do not properly monitor the system failure trends over time. Therefore, systems often exhibit similar failures.

2.2 Issues and challenges of current troubleshooting methodologies

All engineering systems were designed to achieve their goals. Events that fail to achieve their goals (i.e. system failures) in such a system can be attributed to an insufficient design. As Turner and Pidgeon pointed out, a system failure may be defined as a characteristic of subsystems that do not contribute to the fulfilment goal of the supersystem. Alternatively, a system failure is the "termination of the ability of an item to perform its required function" [9]. The predominant technology in current IT troubleshooting is a predefined goal-oriented model. In this model, Van Gigch [10] highlighted the main shortcomings of a system improvement as follows:

- Engineers tend to try to find malfunctions inside the system boundary.

- Engineers tend to focus on returning the system to normal. Long-term improvements cannot be achieved through operational improvements.
- Engineers tend to have incorrect and obsolete assumptions and goals. In most organisations, the formulation of assumptions and goals is not explicit. In this context, improvements to fostering systems are senseless.
- Engineers tend to act as "planner followers" rather than as "planner leaders". In a system design concept, the planner must be a leader planning to influence trends, rather than a follower planning to satisfy trends.

Explanations of system failures in terms of a reductionist approach (i.e. an event chain of actions and errors) are not useful for improving the system designs [11], [12]. In addition, Perrow [13] argued that a conventional engineering approach to ensure safety by building more warnings and safeguards fails because the system complexity makes failures inevitable. The following four key features have commonly been pointed out as limitations of current troubleshooting methodologies in IT system environments.

(1) Most methodologies have a reductionist perspective. This makes it difficult to understand the real meaning of the countermeasures, whether they are effective or tentative.
(2) The current mainstream troubleshooting approach applies a cause–effect analysis (or event chain analysis) to determine the real root causes. FMEA or event trees utilise forward sequences, and FTA or fault trees utilise backward sequences [3], [4].
(3) The speed of intense technological advances creates critical misunderstandings among stakeholders. Current methodologies cannot properly manage the disjunction among stakeholders.
(4) An improvement of the deviation from the operating norm is bound to fail, and as Van Gigch [10] pointed out, the treatment of a system problem is bound to fail when improving the operation of existing systems.

To summarise these four points, the current methodology focuses on the following three issues:

- The system does not meet the pre-defined goals.
- The system produces no predicted results.
- The system does not work as expected at the design phase.

As the basic assumption of improvement, the goal and operating norm are static and predetermined in the design phase and are based on hard-systems thinking.

These four key features and three issues hinder the examination of system failures from a holistic standpoint, making it difficult to manage the soft, systemic, emergent, and dynamic aspects of such failures.

Based on the discussion described in this section, there are four major shortcomings of the current methodologies.

1. A lack of a methodology covering the world views of multiple stakeholders (Section 2.2, shortcoming (3)).
2. A lack of a methodology covering emergent failures (Sections 2.1 and 2.2, shortcomings (2) and (4)).
3. A lack of a methodology covering a holistic view of system failures rather than a reductionistic view (Section 2.2, shortcoming (1)).
4. A lack of a methodology to monitor system failure trends over time (Section 2.1).

In this study, a new meta-methodology (Section 3) called SOSF is proposed as a countermeasure to the first, second, and third shortcomings. In Section 4, a risk quantification/visualisation method is introduced to address the fourth shortcoming.

3 PROPOSAL OF NEW META METHODOLOGY

SOSF promotes double-loop learning to overcome the first, second, and third shortcomings mentioned in the previous section. Double-loop learning is indispensable for reflecting whether operating norms (i.e. mental models) are effective [14]–[16]. A meta-methodology was used to transform the mental models [10]–[13], as described in Section 2.2. The system of system methodologies (SOSM) proposed by Jackson [17] is a typical and widely recognised meta-methodology. Its main features are the use of a meta-systemic approach (soft system thinking to promote double-loop learning) and complementarianism by encompassing multiple paradigms depending on the state of the problem. FTA [2], FMEA [1], and other analysis types, as discussed in Section 2.2, belong to the simple unitary domain in SOSM.

To overcome the first shortcoming (i.e. covering multiple worldviews of stakeholders), SOSF is designed on the SOSM base that covers multiple stakeholders (i.e. the plural domain in SOSM). In particular, SOSF was developed by placing each type of failure from the system failure taxonomy [18] onto a two-dimensional SOSF (left side of Fig. 1). Notably, the recursive and hierarchical features of SOSF depend on the viewpoint of the system. These features form a system as a structural aggregation of subsystems, where each subsystem has its own SOSF. It is therefore important to note the hierarchical and relative structures (i.e. a technology failure may be an evolutionary failure depending on the viewpoint of the subsystem).

Taxonomy of system failures (Van Gigch [30])

Type of failure	Symptom or component
Failure of structure	Levels of recursion & domain/structure
Failure of technology	Hardware/Software
Failure of regulation	Regulation & control/viability/viable regulation
Failure of rationality	Evidence/Rationalities/Structural
Failure of behavior	Syntactics & Semantics/Psychology/Ergonomics/Pragmatics
Failure of evolution	Disjunction of Values/Structure

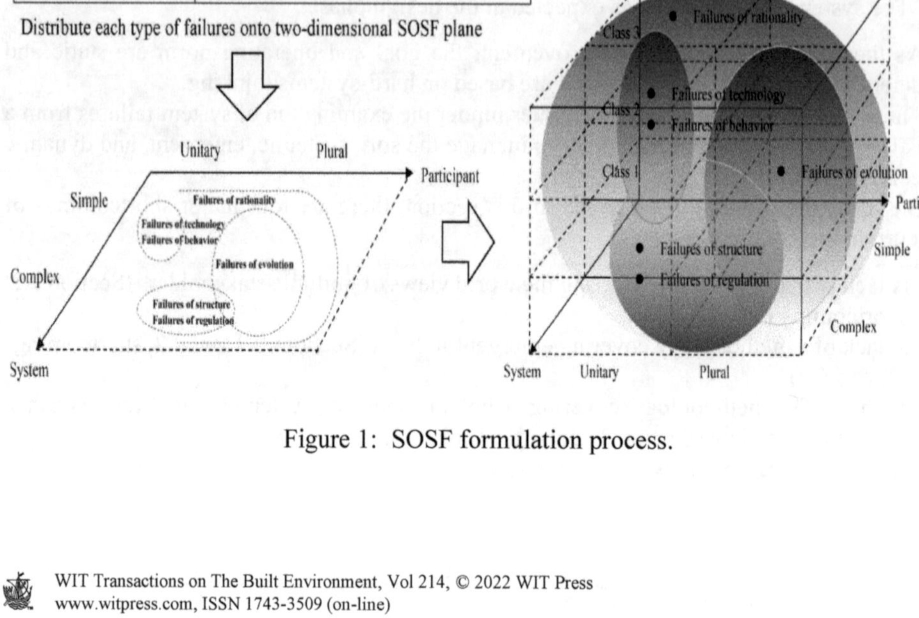

Figure 1: SOSF formulation process.

To overcome the second and third shortcomings (i.e. to cover emergent failures and a holistic view), we introduced a third dimension, namely the failure class. Three failure classes were defined to address the emergent and holistic aspects of a system failure. As Nakamura and Kijima [19] pointed out, failures are classified according to the following criteria.

- Class 1 (deviance failure): The root causes are within the system boundary, and conventional troubleshooting techniques are applicable and effective.
- Class 2 (interface failure): The root causes are outside the system boundary but are predictable at the design phase.
- Class 3 (foresight failure): The root causes are outside the system boundary and are unpredictable during the design phase.

The right-hand side of Fig. 1 (i.e. three-dimensional SOSF) is an expansion of the two-dimensional SOSF (left-hand side of Fig. 1) with the addition of the system failure dimension (i.e. three failure classes).

4 RISK QUANTIFICATION/VISUALISATION METHODOLOGY

The IC chart and closed-code metrics are introduced in this section, including how topological metrics are introduced in the SOSF space to quantitatively monitor the system risk over time.

4.1 Normal accident theory and IC chart

The normal accident theory divides the analysed system into two axes (the coupling axis, i.e. the degree to which the components combine, and the interaction axis, i.e. the degree to which the analysed system interacts with the external environment). The plane represented by these two axes is called an IC chart. An IC chart was first proposed by Perrow [13], and various social systems have been qualitatively laid onto the IC chart plane, as shown in Fig. 2. The IC chart allows a qualitative analysis of the system as two independent variables. Several failures occur, both of which may have a low impact. However, the complex and unexpected interactions of these low-impact failures become fatal owing to a cascading series of minor failures before the safety equipment or alarm comes into effect, which is known as a normal accident against a single-point failure.

To introduce a metric into the IC chart, this qualitative argument should be confirmed using this quantitative measure. As Perrow [13] points out, there is currently no reliable way to measure these two interaction and coupling variables. Therefore, a quantitative discussion is crucial. By introducing this metric into the IC chart, a numerical discussion (change in the position of the failure and time series) can be applied using the IC chart. To address this quantitative issue, the next section explains the method for introducing a metric into the IC chart. A closed-code system of an object system failure is introduced as a metric. If an appropriate topology is added to the IC chart, then a quantified visualisation is possible, enabling us to understand the direction of change in the target system, and thus apply effective countermeasures. The use of this metric enables the safety and quality of a target system to be monitored quantitatively over time.

4.2 Close-code metrics as an example taxonomy of system failures

To introduce a metric into the IC chart described in Section 4.1, we focused on a closed-code system. A close-code system is a type of table that classifies the causes of failures in each

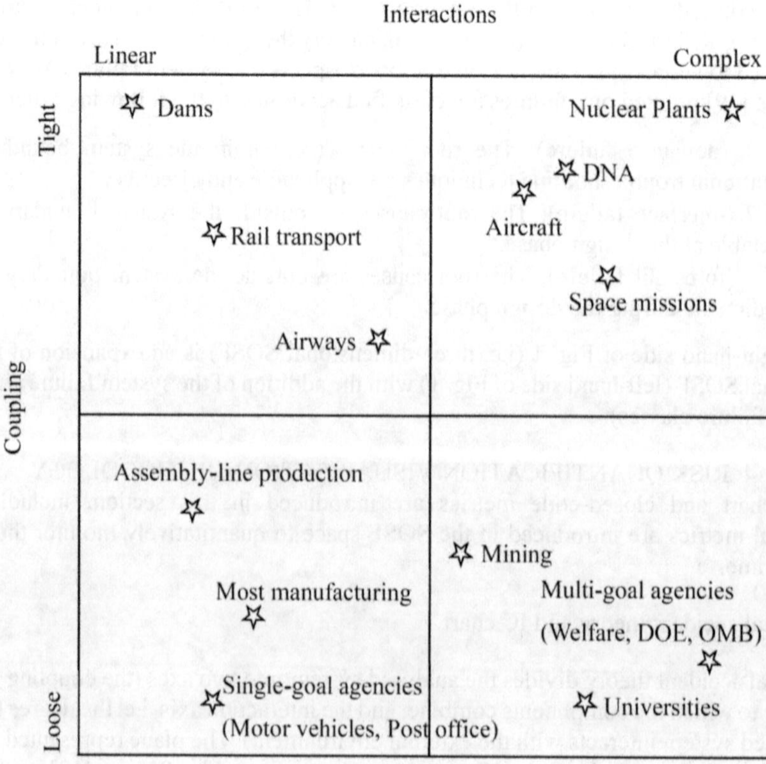

Figure 2: IC chart [13].

industry sector. A close-code is a failure root case classification taxonomy (e.g. hardware, software, or human error) used in a system.

Although the closed-code system varies by system and industry, it is classified as a closed-code matrix with two dimensions. The first dimension consists of phases for creating an object system (i.e. design, configuration, and operation in a time sequence), and the second dimension is the nature of the stakeholders (i.e. simple or complex) responsible for the system failures. A closed-code system is a filter for the root cause of a system failure, and is an example taxonomy of system failures [10], [18]. If a taxonomy of system failures is regarded as a generalised closed-code system, a different closed-code system can be standardised for each industry. This idea enables a quantitative discussion of the nature and changes in the time-series of failures within an industry, enabling comparisons between industries.

Table 1 summarizes an example of mapping a close-code system onto a close-code matrix for the ICT industry, and the relationship between a close-code matrix and an IC chart is clarified. The horizontal (vertical) axis of the close-code matrix corresponds to the interaction (coupling) axis of the IC chart. The coordinates of the close-code matrix in Table 1 are represented by (m, n), where m(n) is the number of horizontal row (vertical column) (i.e. (m = 1: design; m = 2: configuration; m = 3: operation) and (n = 1: simple; n = 2: complex)). For example, (2, 2) indicates configuration-complex area. This (m, n) notation enables risk factors to be quantified. The next section introduces the metric derived from a closed-code matrix used in an SOSF space.

Table 1: Example close-code matrixes in ICT industry.

	Close codes	1 (Design) Failure of technology and structure	2 (Configuration) Failure of regulation	3 (Operation) Failure of behaviour and evolution
		Failure of rationality, evolution		
1 (Simple)	A (Hardware)	A (1–5)[*1]		A(B)[*2]
	B (Behaviours)		B (A–D, F)[*6]	B(E)[*6]
	P (Obsolete)		P	
2 (Complex)	A (Hardware)	A (6)[*3]		A(U)[*4]
	B (Behaviours)			B(G)[*6]
	N (Future plan)	N		

Legend: (Causes A and B have subcategories)
[*1] A(1)–A(5): CPU, memory, channel, power, and disk failure, respectively.[*2] A(B): hardware setup mistake. [*3] A(6): other IO. [*4]A(U): unknown causes.[*5]B(A)–B(G): network setup, IO setup, parameter setup, installation, operation, application coding mistake, and other mistakes, respectively.

4.3 Introduction of metric into SOSF space

In Section 3, SOSF was introduced, and in the previous section, a metric capable of a quantitative discussion when using a closed code system was described. Here, the metric detailed in the previous section is introduced into SOSF. Consequently, SOSF is transformed into a topological space, enabling a quantitative discussion. Each failure in an SOSF space is represented by a point representing the system risk location (SRL). The risk of the target system is represented by a three-tuple. This can be expressed as an SRL within an SOSF space. An SRL is represented by three-dimensional coordinates (X, Y, Z) in an SOSF space, where X (Y, Z) represents the system interaction (system coupling and annual call rate (ACR)). ACR is the ratio of incidents per unit of shipment each year. Isomorphism occurs among the two-dimensional SOSF (left side of Fig. 1), closed-code matrix, and IC chart. The isomorphs of these three perspectives with the component attributes are shown in Fig. 3.

There are four steps used to introduce these metrics into an SOSF space, as shown in Fig. 4. The first step (Fig. 4(a)) defines the system failure group at any arbitrary time.

Fig. 4(b) shows that β is the area inside α; therefore, β is obtained by dividing the number of system failures in the (3, 2) area by the total number of system failures. The X–Y axes in Fig. 4(b) correspond to the interaction-coupling axes. The quantification of risk factors is achieved using the (m, n) notation in the close-code matrix.

We define γ as shown in Fig. 4(c). The complex and loose risk factors of an object system are represented by $\gamma = (\alpha, \beta)$, which is the quantitative coordinate point in the IC chart.

Fig. 4(d) provides a detailed explanation of $\gamma = (\alpha, \beta)$ in an IC chart for the system failure group at any arbitrary time. Adding a new dimension (i.e. the Z-axis representing ACR) to γ produces the SRL (α, β, ACR). Here, ACR is the frequency of system failures and should therefore be incorporated into the system failure metric.

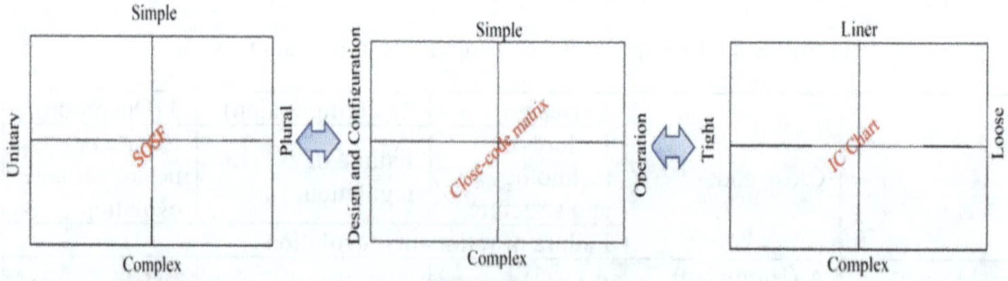

Axis	SOSF		Taxonomy of system failures (Van Gigch)		Example; Close code matrix		IC chart		Metric SOSF SRL (X, Y, Z)
X	Stakeholders	Unitary	Failure of:	Technology, Structure, Regulation, Rationality and Behavior	System Creation Phase	1 (Design) and 2 (Configuration)	Interaction	Linear	N/A
		Plural		Rationality and Evolution		3 (Operation)		Complex	Interaction Metrics (α): (3, n)/all incidents
Y	System feature	Simple	Failure of:	Technology, Behavior, Rationality and Evolution	System feature	Simple	Coupling	Tight	N/A
		Complex		Structure, Regulation, Rationality and Evolution		Complex		Loose	Coupling Metrics (β): (3,2)/all incidents
Z	Failure class (i.e., Class1,2 and 3)		N/A		N/A		N/A		ACR

Figure 3: Isomorphic structure of three perspectives and its component attributes.

In Sections 3 and 4, a topological SOSF was introduced, and a system failure can now be discussed numerically. To confirm its effectiveness in the next section, a new methodology is applied to the ICT system shown in Fig. 5. The virtualised ICT systems in Fig. 5 are mainly composed of three technologies (operating systems, networks, and virtualisation platform products). Virtualised ICT systems and other IOTs (i.e. internet-based information architectural devices) are also complexly connected throughout the networks.

5 APPLICATION

As described in the Introduction, ICT technologies are drastically changing. Complexly connected ICT systems are considered suitable for verifying the effectiveness of the proposed methodology. An overview of the application target system is shown in Fig. 5.

The SRL transition for virtualised ICT systems over a 3 year period is shown in Table 2, and SRL (α, β) was calculated based on the incidents that occurred in the corresponding system components (i.e. OS, virtual platform, and network). Every system failure was identified and gathered from incidents reported by the field operation group and analysed quantitatively by an ICT company. A closed-code matrix (Table 1) was used to formulate a metric within the SOSF space. According to Table 2, the network and virtualisation platforms move towards a complex-loose direction with an increase in the ACR. The OS moves towards a linear-tight direction with a decrease in the ACR.

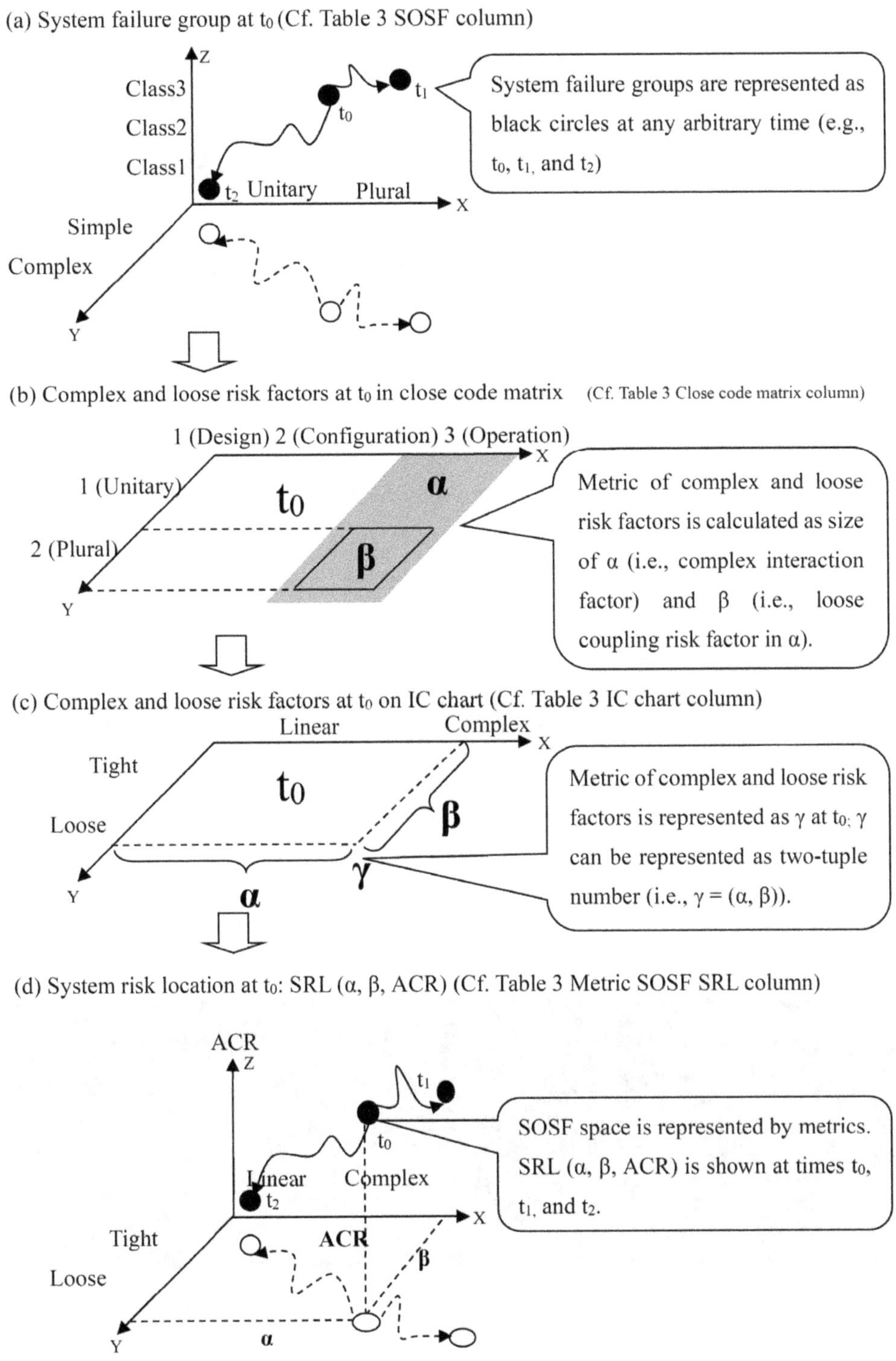

Figure 4: Detailed diagram of metric generation.

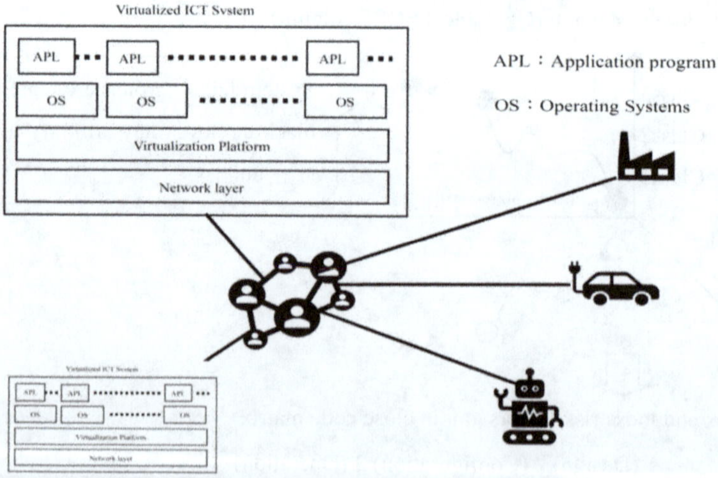

Figure 5: Overview of complexly connected ICT systems.

Table 2: SRL transition.

	2016			2017			2018		
	α	β	ACR	α	β	ACR	α	β	ACR
OS	19.3	19.1	15.5	19.5	19.2	20.6	18.1	18.0	17.6
Virtual	16.7	16.4	31.2	18.0	16.8	39.6	20.3	19.2	41.7
Network	12.6	8.9	44.6	13.0	8.7	57.2	15.9	12.1	52.0

The SRL of the network and virtualisation platform migrates towards the complex-loose direction, and the OS migrates towards the linear-tight direction. The SRL trajectory for each product is shown in Fig. 6. A brief explanation of the key application results is presented below. The network and virtualisation platform migration trends are attributed to changes in the external environment, whereas the OS migration trend is attributed to improved quality and reliability.

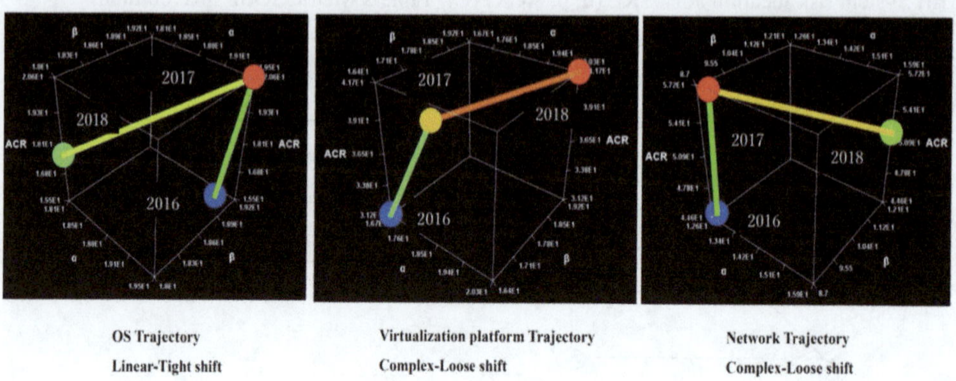

Figure 6: SRL trajectory of each product.

6 CONCLUSION

In this paper, a risk quantification/visualisation methodology for an SOSF space was proposed and applied to complexly connected ICT systems. As an application, the risk over time can be clearly visualised using the quantified SRL within the SOSF space.

The findings of this application are as follows:

1. The SRL of the network and virtualisation platforms has been shifting towards a complex-loose domain from a linear-tight domain. In 2018, the ACR increased from its value in 2017.
2. The SRL of the OS platform shifted towards a linear-tight domain from a complex-loose domain. The ACR in 2018 decreased from its value in 2017.

The first result is believed to be due to the diversity of the stakeholders and the complexity of the technology in relation to networks and virtual platforms, particularly complex shifts that deteriorate the risk over time. This change supports Gartner's analysis, as described in the introduction. This loose shift is believed to be due to system redundancy (such as a network or server duplex).

The second result is believed to be a continuous improvement in quality from a relatively small number of OS vendors in comparison to that of network and virtualisation development vendors, and the speed of change in OS technologies is relatively slow compared to that of networking and virtualisation.

For greater safety and higher reliability, the application results suggest the following. If complex shifts are inevitable owing to technological changes, measures such as the introduction of redundancy enhancement equipment contribute towards a movement away from a catastrophic outcome. In other words, measures such as avoiding complex shifts can enhance the reliability and safety of ICT systems. The results of this application in ICT systems suggest that it may be effective for other social systems as well.

Through its application in an ICT system, the proposed methodology shows the effectiveness of quantitatively monitoring the level of risk over time.

Further research is required to expand this approach to other industries with various close-code matrices. This method will lead to a refinement of the proposed methodology and thus contribute to enhancing the safety and security of our society as a whole.

REFERENCES

[1] Nakamura, T. & Kijima, K., System of system failure: Meta methodology to prevent system failures. *System of Systems*, ed. A.V. Gheorghe, IntechOpen: London, pp. 31–56, 2012.

[2] Cooney, M., Gartner: 10 key IT trends for 2012. https://www.networkworld.com/article/2220899/gartner--10-key-it-trends-for-2012.html. Accessed on: 17 Mar. 2022.

[3] IEC 60812:2018, Procedure for failure mode and effect analysis (FMEA and FMECA). https://webstore.iec.ch/publication/26359. Accessed on: 17 Mar. 2022.

[4] IEC 61025:2006, Fault tree analysis (FTA). https://webstore.iec.ch/publication/4311. Accessed on: 17 Mar. 2022.

[5] Bell, T.E., Special report: Managing Murphy's law: Engineering a minimum-risk system. *IEEE Spectrum*, pp. 24–57, 1989.

[6] Wang, J.X. & Roush, M.L., *What Every Engineer Should Know About Risk Engineering and Management*, Marcel Dekker Inc., 2000.

[7] Beroggi, G.E.G. & Wallace, W.A., Operational risk management: A new paradigm for decision making. *IEEE Transactions on Systems, Man and Cybernetics*, **24**(10), pp. 1450–1457, 1994.

[8] Checkland, P. & Holwell, S., *Information, Systems and Information Systems: Making Sense of the Field*, John Wiley, 1998.
[9] Turner, B.A. & Pidgeon, N.F., *Man-Made Disasters*, 2nd ed., Butterworth-Heinemann: UK, 1997.
[10] Van Gigch, J.P., *System Design Modeling and Metamodeling*, Plenum: New York, 1991.
[11] Rasmussen, J., Risk management in a dynamic society: A modeling problem. *Safety Science*, **27**(2/3), pp. 183–213, 1997.
[12] Leveson, N., A new accident model for engineering safer systems. *Safety Science*, **42**(4), pp. 237–270, 2004.
[13] Perrow, C., *Normal Accidents Living with High-Risk Technologies*, Princeton Paperbacks: New York, 1999.
[14] Morgan, G., *Images of Organization*, New edition, SAGE: California, 1997.
[15] Argyris, C. & Schoen, D., *Organizational Learning II*, Addison Wesley: Massachusetts, 1996.
[16] Senge, P., *The Fifth Discipline: The Art and Practice of the Learning Organization*, 1st ed., Doubleday: New York, 1990.
[17] Jackson, M., *System Thinking: Creative Holism for Managers*, John Wiley, 2003.
[18] Van Gigch, J.P., Modeling, metamodeling, and taxonomy of system failures. *IEEE Transactions on Reliability*, **R-35**(2), pp. 131–136, 1986.
[19] Nakamura, T. & Kijima, K., System of system failures: Meta methodology for IT engineering safety. *Systems Research and Behavioural Science*, **26**(1), pp. 29–47, 2009.

EFFECT OF DIFFERENT NANOMATERIAL ADDITIONS IN CLAY-BASED COMPOSITES

IVAN VRDOLJAK[1], IVANA MILIČEVIĆ[1] & SLAVKO RUPČIĆ[2]
[1]Faculty of Civil Engineering and Architecture, Josip Juraj Strossmayer University of Osijek, Croatia
[2]Faculty of Electrical Engineering, Computing, and Information Technologies, Josip Juraj Strossmayer University of Osijek, Croatia

ABSTRACT

In this study, electromagnetic (EM) radiation shielding of a clay composite with three different admixtures was tested. Because of the rapid development of new modern technologies, human exposure to non-ionizing EM radiation has grown exponentially. This has aroused the interest of scientists to answer the question of how long-term exposure to EM radiation is harmful to human health. Due to potential risks to human health and an increasing number of studies that have concluded that long-term EM radiation can be harmful to health, the WHO has classified the radiofrequency band (from 300 kHz to 300 GHz) as a potential cause of cancerous diseases. The frequency span 1.5–6 GHz was tested. Admixtures used as a partial replacement in mass percentage are fly ash, titanium dioxide and zinc ferrite. The percentage replacement was 5%. The results showed the lowest transmission in specimens with titanium dioxide addition. The lowest transmission in regards to the reference sample was about −10 dB. This research was done with the aim of production of load-bearing bricks with high protection against EM radiation for the construction of healthy buildings.

Keywords: clay, fly ash, titanium dioxide, zinc ferrite, electromagnetic radiation shielding.

1 INTRODUCTION

Electromagnetic (EM) radiation can be defined as the transmission of energy that travels throughout the medium in the shape of electric and magnetic fields. EM radiation travels at the speed of light. Probably the "most popular" type of EM radiation is light. Depending on the frequency, electromagnetic radiation includes radio waves, microwaves, infrared, light, ultraviolet, X-rays and gamma rays. All these waves belong to the electromagnetic spectrum. As the frequency of a wave is higher, it transmits higher energy. At one point, that energy becomes so high, it can ionise atoms by detaching electrons from them. This type of radiation is called ionising radiation. Exposure to such types of radiation is not frequent. People are often exposed to low intensity ionising radiation in hospitals due to examinations. However, exposure to non-ionising radiation is much more common. People are every day surrounded with a mixture of electric and magnetic fields. Non-ionising radiation comes from various devices such as mobile telephony systems (3G, 4G, 5G), wi-fi, wireless devices, radars, home appliances, among others. Each of these sources operates at different frequencies. Short time exposure to higher amounts of non-ionising radiation can cause decreased concentration, sleep problems, mental disorders and stress. While scientists' opinions on short-term exposure to non-ionising radiation are similar, opinions on the consequences of long-term exposure to non-ionising radiation are divided. One part of scientists believe that long-term exposure does not have a negative impact on human health, while other part of scientists believe long-term exposure to non-ionising radiation may result in an increased risk of developing cancer. Khurana et al. [1] compared the risk of developing neurobehavioral symptoms or cancer in populations living at distances < 500 m from base stations in available studies. Authors reported an increased prevalence of adverse neurobehavioral symptoms or cancer in populations living at distances < 500 m from base stations found in 80% of the available studies. Calvente et al. [2] published a comprehensive review of the literature on

www.witpress.com, ISSN 1743-3509 (on-line)
doi:10.2495/SSR220131

the subject, studying research from 1979 to 2008. The results of the literature review showed that most research in this period found an association between childhood leukaemia and exposure to electromagnetic radiation, especially when exposed to very low frequencies. Because of these growing concerns, the IARC (International Agency for Research on Cancer) 2001 classified very low frequencies as potentially carcinogenic [3]. Ten years later, in 2011, the same agency added radio frequencies (from 300 kHz to 300 GHz) to the group of potentially carcinogenic samples [4]. The IARC divides potential cancer agents into three groups: (1) they are carcinogenic; (2) they are probably carcinogenic; and (3) potentially carcinogenic. Due to insufficient research, but growing concerns, exposure to non-ionising radiation belongs to the "weakest", the third category. There are three most common ways of reducing our exposure to non-ionising radiation: First, limiting time spent in rooms with high levels of electromagnetic radiation, second, gaining distance – just as the heat from a fire decreases as a person moves away, the radiation dose decreases dramatically as you increase the distance from the source and third, the use of walls or barriers that have the ability to protect against EM radiation which can significantly reduce the spread and exposure to electromagnetic radiation. The first two ways cannot be considered as long-term solution, while the use of barriers or walls in rooms can significantly reduce long-term exposure to EM radiation. Materials such as clay bricks cannot be barriers by itself because of the high electrical resistivity that clay has (around 106 Ωm) [5]. This study explored way of improving EM shielding potential of clay composites by adding different additives which have been shown to be effective in concrete. Analysed additives are: fly ash, zinc ferrite ($ZnFe_2O_4$) and titanium dioxide (TiO_2) [6]. The additives percentage in clay composite was 5% by mass. The study analysed the frequencies to which people are most often exposed including: 4G bands (1.80–1.88 GHz, 2.11–2.17 GHz, 2.62–2.69 GHz), 5G band (3.45–3.80 GHz), wi-fi (2.4 GHz) and navigational radar (2.9–3.1 GHz).

2 SPECIMEN PREPARATION

Clay was supplied from a local brick factory, after which it was dried and reduced to the desired granulation. Clay and additives were mixed in a dry state, after which water was applied to achieve desired moisture (around 22 wt.%). For each additive, three specimens were made to achieve more reliable results. Specimens were circular with 15 cm diameter and 2 cm thickness. Clay specimens were moulded and compacted with the Proctor device. After that, specimens were dried at adequate room humidity to avoid any cracking. Dried specimens were exposed to firing at a temperature of 850°C. Fig. 1 shows the heating program.

3 EXPERIMENT SETUP

Transmission measurement was achieved using Anritsu ms2038c Handheld Vector Network Analyser and Spectrum Analyser. The analysed frequency was 1.5–6.0 GHz, where scattering parameters (S-parameters) were measured for providing transmission, reflection and absorption values. As mentioned before, for each additive, three specimens were made, after which the mean value was taken and compared to the other values. There are various standard test methods for measuring shielding efficiency. They can be classified into small-sample testing and large-sample testing. ASTM D4935-18 can be considered representative for small-sample testing. The problem with this method is the specimen's small thickness (around 25 μm) that is not possible to fabricate [7]. On the other hand, standards like IEEE-STD-299 that is a representative large-sample test, prescribe a higher testing area, and the specimen dimensions need to be 2.0 m^2 × 2.0 m^2. For these reasons, an experiment was based

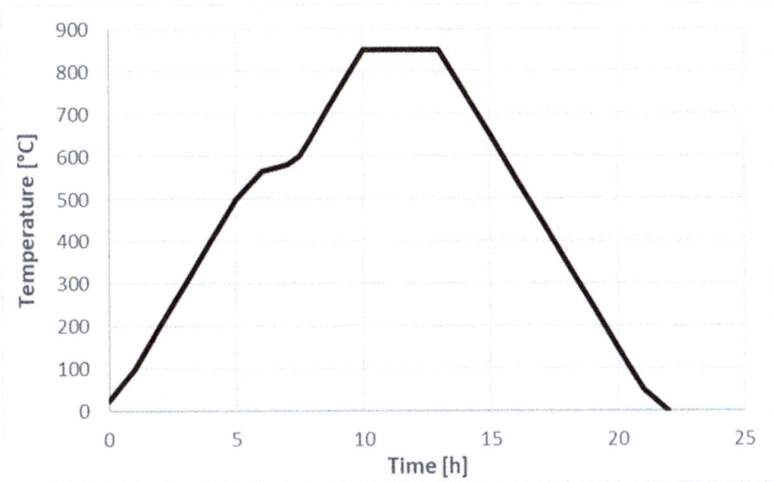

Figure 1: Firing process.

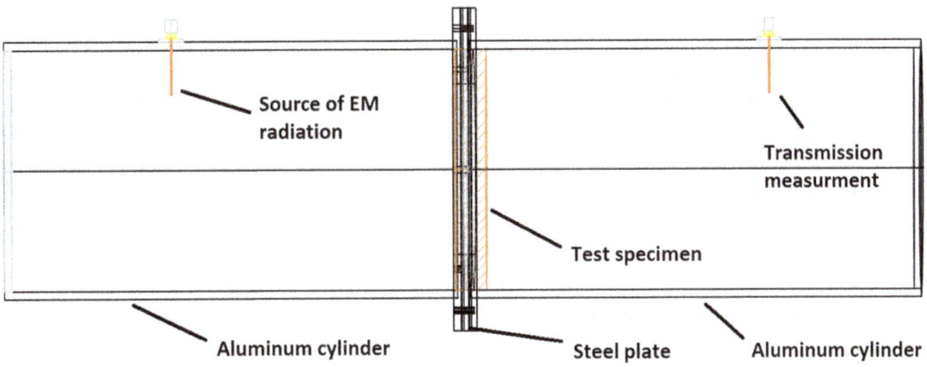

Figure 2: Experiment setup.

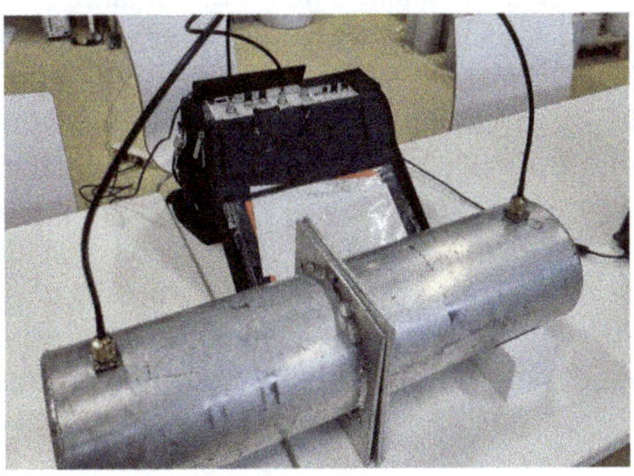

Figure 3: Measuring process.

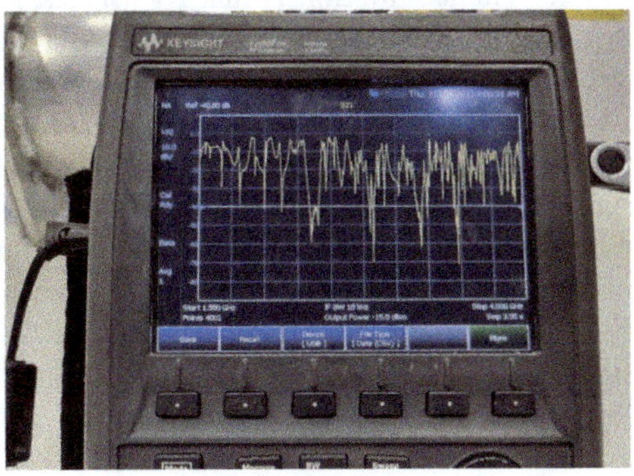

Figure 4: Transmission measurement.

on an enlarged ASTM D4935-19 test. The measurement setup consisted of two aluminium cylinders between which a specimen holder was placed (Fig. 2). To avoid any space between specimen and specimen holder, aluminium foil was placed. Two cylinders and the specimen holder were connected with screws.

4 RESULTS

Transmission, absorption and reflection results were measured for a frequency span 1.5–6.0 GHz (Fig. 4). At these frequencies, most of the fixed radiation sources are. Special attention was given to frequencies at which mobile telephony systems operate, wi-fi networks and navigational radar: 4G bands (1.80–1.88 GHz, 2.11–2.17 GHz, 2.62–2.69 GHz), 5G band (3.45–3.80 GHz), wi-fi (2.4 GHz) and navigational radar (2.9–3.1 GHz). Values are presented using S-parameters. S_{21} parameter displays the transmission. The transmission shows the difference in the level of electromagnetic radiation before and after passing through the sample. The lower the values of the curves in the diagrams then the lower the transmission is. Measurement results are shown in Fig. 5.

From Fig. 5(a), (b) and (e), it can be concluded that adding zinc ferrite, titanium dioxide or fly ash to clay did not reduce transmission in first two 4G frequency spans or wi-fi frequencies. Although first two frequency spans at which 4G operate did not show any transmission reduction, third frequency span (2.62–2.69 GHz) has shown significant transmission reduction in specimens where titanium dioxide was used. Transmission reduction was around −10 dB. In 5G band (3.45–5.80 GHz) specimens with titanium dioxide again have proven to be most efficient for transmission reduction. Maximum reduction was −6.7 dB at 3.6 GHz. Highest reduction in transmission at frequencies where navigational radar operates have again been shown in specimens with titanium dioxide. Transmission reduction was reported throughout the entire range from 2.9–3.1 GHz. Maximum reduction was measured at 3 GHz and it was 8.3 dB. Specimens with fly ash or zinc ferrite did not show any significant reduction in any of the analyzed frequency ranges.

5 CONCLUSION

Due to growing concerns about the potentially harmful long-term effects of non-ionizing radiation, more and more research is investigating how to reduce our exposure most effectively to radiation. This study analyzed the electromagnetic shielding capabilities of the clay composites for the frequency range of 1.5–6 GHz. Admixtures that were used as partial clay replacement were: fly ash, titanium dioxide and zinc ferrite. Clay composite specimens containing fly ash or zinc ferrite did not show significant change relative to the reference sample. The assumption is that fly ash does not have a sufficient amount of hematite in its chemical composition (most often up to 10%), which increase conductivity of the material and lead to reduced transmission of EM radiation. Despite the fact that zinc ferrite has good magnetic properties, the results showed minimal improvements in electromagnetic shielding (only a few dB). The results are in line with those from other literature, where a significant improvement in protection was shown only when a 50% replacement of zinc ferrite was used in the mortar. Clay composites with titanium dioxide have shown highest potential for electromagnetic shielding. The highest transmission reduction measured was around 9 dB at 4G frequency span 2.62–2.69 GHz. These results show the possibility and potential of using titanium dioxide in real clay bricks that could provide significant protection against electromagnetic radiation.

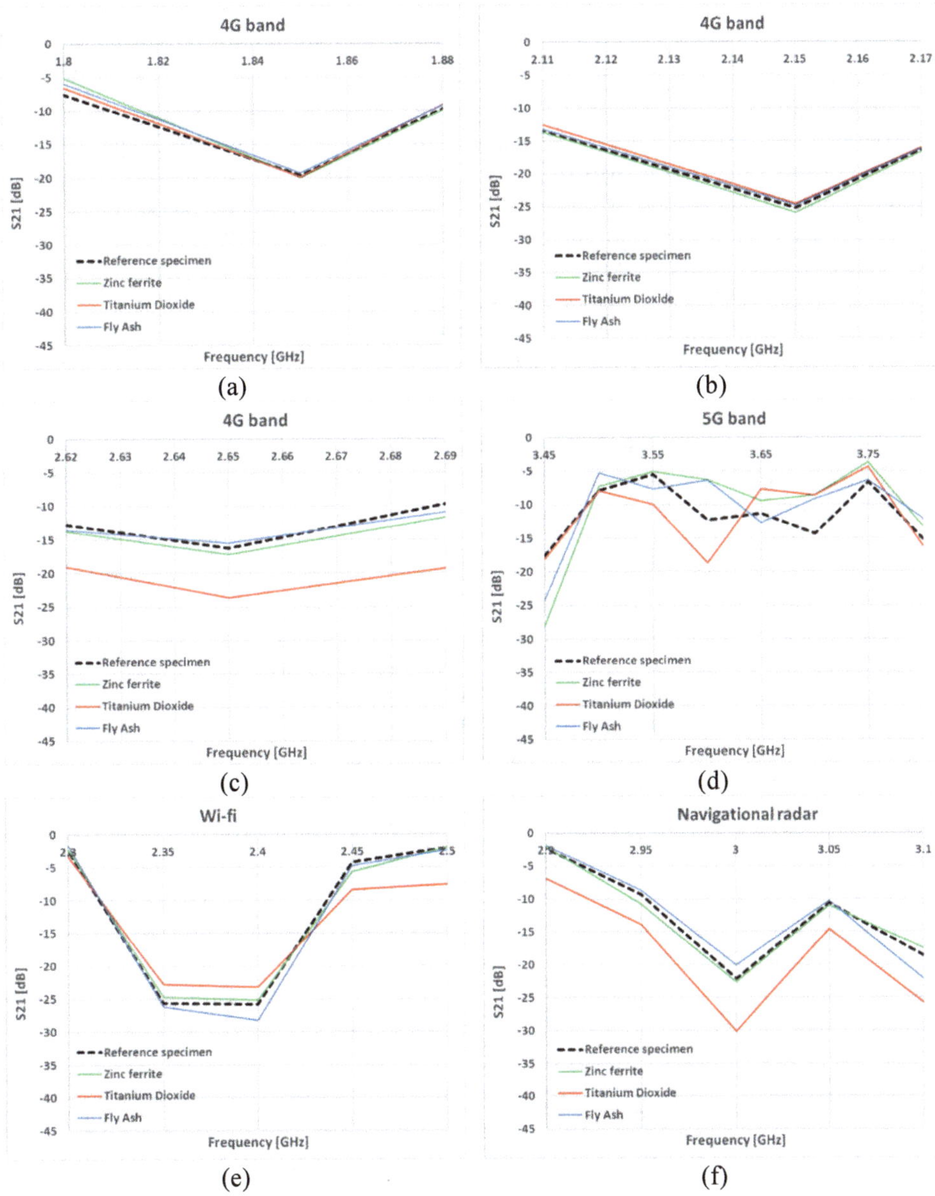

Figure 5: Measured S_{21} parameter of clay composite samples for (a) 4G, 1.80–1.88 GHz; (b) 4G, 2.11–2.17 GHz; (c) 4G, 2.62–2.69 GHz; (d) 5G, 3.45–3.80 GHz; (e) Wi-fi, 2.4 GHz; and (f) Navigational radar, 2.9–3.1 GHz.

ACKNOWLEDGEMENT

This work has been supported by the Operational Program on Competitiveness and Cohesion 2014–2020; Investing in science and innovation KK.01.1.1.04: Development and application of advanced building materials for the construction of healthy buildings: protection against non-ionizing radiation, ref. no. KK.01.1.1.04.0105.

REFERENCES

[1] Khurana, V.G., Hardell, L., Everaert, J., Bortkiewicz, A., Carlberg, M. & Ahonen, M., Epidemiological evidence for a health risk from mobile phone base stations. *International Journal of Occupational and Environmental Health*, **16**(3), pp. 263–267, 2010. DOI: 10.1179/oeh.2010.16.3.263.

[2] Calvente, I., Fernandez, M.F., Villalba, J., Olea, N. & Nuñez, M.I., Exposure to electromagnetic fields (non-ionizing radiation) and its relationship with childhood leukemia: A systematic review. *Science of the Total Environment*, **408**(16), pp. 3062–3069, 2010. DOI: 10.1016/j.scitotenv.2010.03.039.

[3] Repacholi, M.H., An overview of WHO's EMF project and the health effects of EMF exposure. *Proceedings of the International Conference on Non-Ionizing Radiation at UNITEN (ICNIR 2003) Electromagnetic Fields and Our Health*, October, Kuala Lumpur, Malaysia, pp. 1–21, 2003.

[4] IARC & World Health Organization, IARC classifies radiofrequency electromagnetic fields as possibly carcinogenic to humans. Press Release No. 208, 2011. http://www.iarc.fr/en/media-centre/pr/2011/pdfs/pr208_E.pdf.

[5] Kuranchie, F.A., Shukla, S.K. & Habibi, D., Utilisation of iron ore mine tailings for the production of geopolymer bricks. *International Journal of Mining, Reclamation and Environment*, **30**(2), pp. 92–114, 2016. DOI: 10.1080/17480930.2014.993834.

[6] Vrdoljak, I., Varevac, D., Miličević, I. & Čolak, S., Concrete-based composites with the potential for effective protection against electromagnetic radiation: A literature review. *Construction and Building Materials*, **326**, 126919, 2022. DOI: 10.1016/j.conbuildmat.2022.126919.

[7] Jung, M., Lee, Y.S. & Hong, S.G., Effect of incident area size on estimation of EMI shielding effectiveness for ultra-high performance concrete with carbon nanotubes. *IEEE Access*, **7**, pp. 183105–183117, 2019. DOI: 10.1109/ACCESS.2019.2958633.

SAFETY CHALLENGES WHEN MANAGING SHIFT WORK IN THE SWEDISH FOREST INDUSTRY: SOLUTIONS BASED ON MORE THAN 700 YEARS OF WORK EXPERIENCE

PIA ULVENBLAD & HENRIK BARTH
Academy of Business, Innovation and Sustainability, Halmstad University, Sweden

ABSTRACT

The purpose of this study is to explore the safety challenges managing shift work in the Swedish forest industry. There are several factors that can affect safety in working life. In general, previous research shows that managerial and leadership behaviours affect safety and well-being regardless of the type of industry. A leadership with a focus on safety affects the safety climate and the outcome in the form of accidents in an organisation. The interaction between managers and employees regarding safety issues also has a positive effect on safety. Leaders who are instead passive regarding safety issues have significant negative effects on safety, contribute to an increased number of accidents and reduce safety-related behaviour in the organisation. This means that employees with passive leaders themselves become less interested in engaging in safety activities. Shift work has also been shown to affect the health and safety of employees. For example, studies show that higher sickness absence is associated with three-shift rotation compared to two-shift rotation and that fatigue and insomnia may be an effect of shift work. Further, research also shows a link between fatigue and a higher frequency of accidents. In this paper, we use a co-creation approach to capture the safety challenges related to shift work. The overall methodological approach in this study is action-oriented research. The empirical data is collected with interviews and focus groups which altogether include 56 respondents with more than 700 years of joint work experiences from shift work in the forest industry. Preliminary results show the importance of creating a safety culture where both the management and the employees take active part and responsibility in solutions. The managers have the formal responsibility, including law and regulation related to safety aspects. The employees in turn have the responsibility for their own choices and behaviours not least related to food, sleep, and training.

Keywords: safety challenges, shift work, forest industry.

1 INTRODUCTION

The purpose of this study is to explore the safety challenges managing shift work in the Swedish forest industry. This is important for several reasons. The forest industry is one of the industries that have a high proportion of workplace accidents. In Sweden the forest industry has had the most workplace incidents with a fatal outcome during the last decade [1]. The Swedish government has formulated an updated work environment strategy for the future for the period 2021–2025 [2]. The Government states that the occupational injuries entail severe financial consequences at the individual and societal level as well as personal suffering for those affected. The work environment strategy for 2021–2025 focuses on the sub-goals (i) a sustainable working life – everyone must be able, strong, and willing to work a full working life; (ii) a healthy working life – working life must contribute to development and well-being; (iii) a secure working life – none should risk life or health due to the job; and (iv) a labour market without crime and cheating. However, the Swedish Work Environment Authority believes that fatal incidents should be greatly reduced by companies developing a good safety culture and functioning systematic work related to the security aspects in the industry.

www.witpress.com, ISSN 1743-3509 (on-line)
doi:10.2495/SSR220141

There are several factors that affects safety in working life. In general, previous research shows that *managerial and leadership behaviours* affect safety and well-being regardless of the type of industry [3]. A leadership with a focus on safety also have an impact on the safety climate and the outcome in the form of accidents in an organisation [4], [5]. Working in businesses with high risks requires leaders to act proactively when it comes to risk management. Previous research shows that in cases where leaders worked to analyse previous events and learned from them, had a positive effect on the management of future crises or accidents [6], [7]. Leaders who are instead passive about security issues have significant negative effects on safety, contribute to an increased number of accidents [8] and reduce safety-related behaviour in the organisation [9]. This means that employees with passive leaders themselves become less interested in engaging in safety activities. Leadership training, not only for formal managers but also for employees, has proven to be positive for both employee well-being and efficiency [10].

The *interaction* between managers and employees regarding security issues also positively impacts security [11]. Previous research in the forest industry in Sweden has shown that if employers and employees have a consensus on e.g., safety culture, it is beneficial towards employee's well-being [12]. Studies from the Finnish forest industry show that employee well-being is also positively affected by being able to participate in change work in relation to one's own work situation [13], [14]. Further, it has positive effects to work together with environmental issues, for example managers, employees, representatives from the unions and health representatives [15].

Shift work has also been shown to affect the health and safety of employees. For example, studies show that higher sickness absence is associated with three-shift rotation compared to two-shift rotation [16] and that fatigue and insomnia may be an effect of shift work [17]. Further, research also shows a link between fatigue and a higher frequency of accidents [18], [19]. Health issues related to shift work also has been shown to include problems with blood pressure [20] and even mortality [21] together with coronary heart disease [22]. In addition, Mylek and Schirmer [23] highlight the importance to broaden employee's wellbeing by implementing workplace strategies that includes not only the physical safety to create a culture of wellbeing in the industry.

In this paper, we use a co-creation approach to capture the safety challenges related to shift work. We also focus specifically on "who can/should do what?" in terms of different perspectives on responsibilities, (i) the company; (ii) the employee; and (iii) the work together.

The paper is structured as follows. Firstly, a frame of reference is outlined with (i) the Swedish context and (ii) theories about co-creation. Secondly, the material and methods used in the paper will be described. Thereafter, the results and discussion are presented. The final part includes conclusions, implications, and future research avenues.

2 FRAMEWORK

2.1 The Swedish context

The Swedish' forest industry includes companies that work with processing the forest into bio-based products such as producers of pulp, paper, board and biofuel, also including sawmills. Of Sweden's total land area 70% is covered by forest and this is about 28 million hectares (of almost 41 million hectares) [24]. As previously mentioned, the forest industry in Sweden has had the most workplace incidents with a fatal outcome during the last decade

[1]. However, these could be reduced by companies developing a good safety culture and functioning systematic work related to the security aspects in the industry.

The law that governs work with work environment in Sweden is primarily the Work Environment Act [25]. According to the Work Environment Act all employees in Sweden have the right to a good work environment. The goal is to reduce risks for accidents and illness in health for the employees. The Act states, for example, that the company/employer and organisation is ultimately responsible for the work environment and thus must lead the work towards a good and better work environment. A special law regulates working hours – the Working Hours Act [26]. Here there is information about how much you can work per day, per week and per year, and to what extent you are entitled to breaks, breaks or meal breaks. The organisational and social aspects of the working environment is focused on in the Law for Organisational and Social Work Environment [27]. The purpose with the regulations is to promote a good working environment and prevent the risk of ill health due to organisational and social conditions in working environment. It relates to (i) the management and control; (ii) communication; (iii) participation, room for manoeuvre; (iv) work division; and (v) requirements, resources, and responsibilities. Beside the law, your work can also be regulated in a collective agreement which is an agreement between the employer and the union. If there is no collective agreement in your workplace only the law is applied.

2.2 Co-creation

In this context it is here argued that a co-creation approach could be of value when identifying risks and developing safe working environment. Co-creation allows involved stakeholders to create value through interactions. Prahalad and Ramaswamy [28] developed the method of DART (dialog, access, risk, and transparency), also used in the exploratory study by Ulvenblad and Barth [15].

Based on the initial work of Prahalad and Ramaswamy, four building blocks provide the pillars for co-creation method of DART namely, *dialog*, *access*, *risk,* and *transparency*. These building blocks challenge in many ways the traditional view managers have taken on issues as labelling laws, disclosure of risks, and transparency of financial statements. Firstly, a *dialog* is very important for the co-creation process to advance as it implies interactivity, deep engagement, and commitment, but also the ability and willingness to change and adapt. *Access* and *transparency* to information for employees as well as employers is necessary for meaningful dialog. More importantly, these building blocks can lead to a unified assessment of the risk-benefits of action and decision of potential HR practices.

Furthermore, the co-creation process is also viewed from a provider and user perspective, or rather employer/manager and employee perspective. This is done to provide information on contribution that is jointly as well as user specific in the co-creation process. Here we use the framework developed by Grönroos and Voima [29], which was originally designed based on buyer–seller integration. In this study we argue that the value creation sphere can be applied in the process of developing safe working environment. In this study we focus on the value creation in the spheres of the employer (including management teams) and employees (Fig. 1).

The firm is responsible for the production process and in the *provider sphere* it produces resources and processes for safety work. In this way the firm facilitates employees value creation. In the *joint sphere*, the role of the employees is twofold: co-producer of resources and processes individually (or jointly between employees) *and* value creator jointly with the

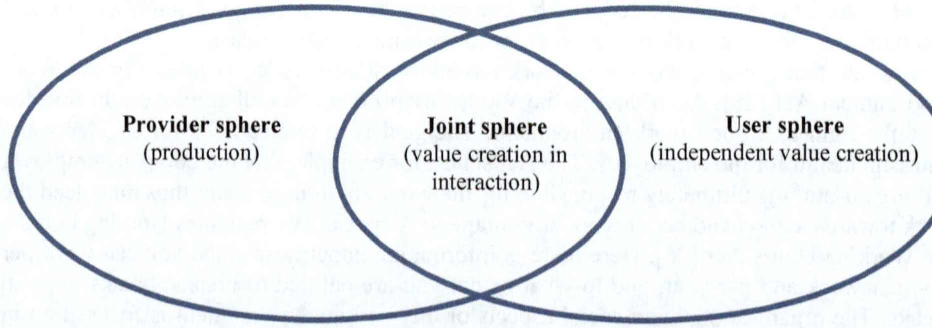

Figure 1: Value creation spheres. Framework designed based on the work of Grönroos and Voima [29].

provider (employer representatives). In direct interactions with the employees, the management team may have an opportunity to engage with the employers' value creation process and take on the role of value co-creator. In the rest of the *employer sphere*, which is closed to the provider, the employers create value as value-in-use independently of the provider. Overall, the questions addressed in the process of developing safe working environment focus on – what can the employer do, what can the employees do, and what can they co-create together to develop safe working environment?

3 MATERIALS AND METHODS

3.1 The research project

This article's empirical data is included in a research and development project. The case included in the study is a company in the forest industry situated in an industrial community in Sweden. The research project, which began in the autumn of 2019, was inspired by the need for development of work environment safety and health in the forest industry. The authors of this article have followed the project from the beginning and conducted a formative evaluation of the project during the process [30]. The project will continue 2020–2022.

The whole project consists of the following three parts: (1) current situation analysis; (2) development work; and (3) final evaluation:

1. *Current situation analysis* – the researchers map the current situation. The company makes an initial action plan based on the needs and interests of both employers, trade unions and occupational health care.
2. *Development work* – the company's own project develops as knowledge and insights grow and are based on an action plan with activities. The researchers are following and supporting the development process.
3. *Final evaluation* – the researchers summarise the project's results together with the project's actors.

The part of the data presented in this paper is included in part 1 entitled *Current situation analysis* and was collected during spring 2021.

3.2 Overall research design

The overall methodological approach in this study is action-oriented research. Participatory action research includes three integrated aspects, namely participation, action, and research [31]. The participation includes several different stakeholders, such as managers as well as representatives from the unions, key persons identified in the process and business health representative. The two authors to this paper have been present in each activity, making it possible to participate actively and at the same time make necessary documentations.

3.3 Data collection and analysis

The empirical data has been collected with interviews and focus groups which altogether include 56 respondents with more than 700 years of joint work experiences from shift work in the forest industry. The respondents were chosen with the help from the company based on two criteria: (1) the respondents should include different stakeholder groups; managers, employees, union, occupational health care etc.; and (2) the group of respondents should include respondents with experiences of shift work. The focus in the interviews were to capture strengths and weaknesses in the work environment as well as possibilities to act and work in new ways. The focus was also especially on (i) what the company can/should do; (ii) what the employee can/should do; and (iii) what areas in the work environment would benefit from working together with different competencies, experiences, and functions.

In the data analysis a content analysis has been employed focusing to compare similarities, and patterns in the data material to make it possible to extract data into the three areas (spheres) of interest and relevance in the study: (i) employer sphere; (ii) employee sphere; and (iii) joint sphere.

4 RESULTS AND DISCUSSION

The results show the importance of creating a safety culture where both the management and the employees take active part and responsibility in solutions. The managers have the formal responsibility including to follow the law related to safety and work environmental aspects. For example, the Work Environment Act [25], the Working Hours Act [26] with regulations focusing working hours and by these possibilities in creating a shift-schedule. Another example is the Law for Organisational and Social Work Environment [27]. The social parts are, among other things, about feeling safe at work and not being exposed to abusive discrimination. It is also about feeling included in the group at the workplace. The employer has challenges regarding for example information between shifts and is working to find solutions by both sending out newsletters and spreading information on regular morning meetings. However, at the morning meetings not all are included because of the shift work. There have also been discussions about delegating responsibility also to the shift-teams and this work is on-going.

The employees in turn have the responsibility for their own choices and behaviours not the least related to, food, sleep, and training. It is obvious that some of the employees with long experience working shift have found their own models to handle both the food during the shift, the sleeping habits related to the shift and the training being able to live a healthy life. We know from previous research that the health issues are of importance related to shift work [20]–[22]. In addition, research has shown a link between fatigue and a higher frequency of accidents [18], [19].

The *employer sphere* focuses on procedures, documentation, and risk analysis (Table 1). The management team also need to address work related information between the different

Table 1: Examples of illustrative quotes for each sphere selected from the 56 interviews.

Employer sphere	Joint sphere	Employees' sphere
Big differences between shift work at daytime and night-time. More to do during daytime. Difficult to make sure that information and communication is provided between the shifts.	We now have a cooperation established; a new co-operation model has been agreed on.	Difficult for the family to understand that I need time to adjust after shift work, I need to rest and sleep…
We act on serious things from the staff survey that we do every year, for example regarding discrimination. Every department receive information and support to make changes.	We have a collegial atmosphere that works very well, "you are more than colleagues", we cover for each other if needed.	Older people have more difficulties doing shift work, they tend to develop diabetes.
We provide information every week on the development and changes to all the employees, like a newsletter. So, everyone knows what is going on.	We have a forum to discuss mutual working aspects – the working environment committee. Cooperation is something we are good at, we are "one big family", good working climate.	It is important that you are fit and healthy, since you are away 48 h every other week. Even if you think that is working for you, you need to be open for signs like depression or anxiety
Usually, the colleagues come to my office, but after the Covid restrictions, we do all the morning meetings online now.	Leisure and work – it feels like the shift teams that have activities together after shifts, feel good.	Do not eat lunch boxes at night, I had to stop drinking coffee "learning by doing".
We have talked a lot about delegating responsibility, those who have the competence, they should work with problem solving, they should get more challenges.	We got a challenge, how do we reduce the workforce we tested and saved 10 in the staff, we have a good negotiating climate.	At night I eat lighter food.
Ok rehabilitation good will in the company, they are good at finding temporary solutions	At every reorganisation we are included, we watch.	One should not change the diet (during the shift)

shifts and provide a system of communication and reporting between the employees. This should include all aspects of safety work environment such as reporting, communication, health, cultural aspects etc.

The *employees sphere* address concerns about health and living conditions when not at work but could lead to risks at work if the employee do not follow these guidelines and recommendations. This also draw the attention to team building and awareness of potential individual problems need to be addressed.

The joint sphere address concerns about developing guidelines for awareness and knowledge when it comes to health issues, support, and system for alerting management team when signs of health or other issues prevail with members in the group.

To summarise, many different challenges need to be addressed when working with safety work environment. Some of these aspects can be related to the management sphere like regulation and law procedures, while other type of risks relates to employees living conditions and recovery-time that need to be addressed and planed for off-work. Furthermore, some of these challenges are best developed jointly to find models and systems to co-create inclusive solutions of how to develop safe working conditions, see Fig. 2.

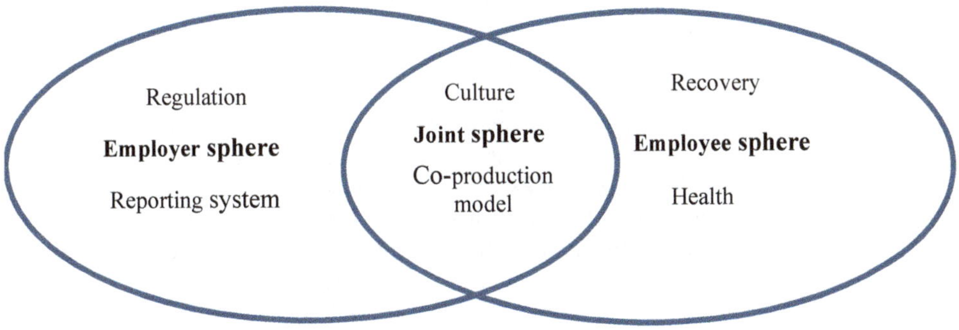

Figure 2: Employer/Joint/Employee spheres in developing a safety culture and work environment.

5 CONCLUSIONS

At least two aspects need to be considered when developing safety work. First, it is important to involve all the members of the organisation. This could be complexed and time-consuming for larger organisation, why small groups could be of interest to work with together with key representatives. This is important as many different features are included and related to safety work. For example, some aspects could be directly linked to work procedures and routines to eliminate risk of accidents. This type of safety procedures could be developed by company representatives like safety managers addressing risk analysis and regulation that need to be followed and reported, which also need to be handled by management team. However, risk is also associated with aspects that is outside of the control and responsibility of the management team, such as the living habits. For example, it is of importance that shift workers have time to recover and, also use that time in a responsible way.

Second, the process of developing safety work needs to be transparent, open, and inclusive. The management team must enforce safety regulations, but this also involves employees when it comes to indirect aspects such as health and living conditions that also could be related to risks and accidents.

There are implications for both employers and employees in co-creating a safety culture and work environment based on inclusion and mutual responsibility. It is also of importance to include stakeholders such as occupational health care to take part with their competence. Future research avenues will benefit from continuing the studies of co-creation in the process of creating a safety culture and work environment. Among other things it is of relevance to capture how to spread the information about working shift in a systematic way from the experienced shift workers to the beginners.

ACKNOWLEDGEMENT

The research was financed by grants from the AFA Insurance, Sweden (AFA Försäkring).

REFERENCES

[1] Swedish Work Environment Authority, Arbetsmiljöverket, Analys av dödsolyckor 2018 och första halvåret 2019. https://www.av.se/globalassets/filer/publikationer/rapporter/analys-av-dods-olyckor-2018-och-forsta-halvan-av-2019-2019-037768.pdf. Accessed on: 1 Mar. 2020.

[2] Government letter, Regeringens skrivelse 2020/21:92. En god arbetsmiljö för framtiden – regeringens arbetsmiljöstrategi 2021–2025.

[3] Kelloway, E.K., Nielsen, K. & Dimoff, J.K. (eds), *Leading to Occupational Health and Safety: How Leadership Behaviours Impact Organizational Safety and Well-Being*, John Wiley, 2017.

[4] Clarke, S., Safety leadership: A meta-analytic review of transformational and transactional leadership styles as antecedents of safety behaviours. *Journal of Occupational and Organizational Psychology*, **86**(1), pp. 22–49, 2013.

[5] Mullen, J.E. & Kelloway, E.K., Safety leadership: A longitudinal study of the effects of transformational leadership on safety outcomes. *Journal of Occupational and Organizational Psychology*, **82**(2), pp. 253–272, 2009.

[6] Carmeli, A. & Schaubroeck, J., Organizational crisis-preparedness: The importance of learning from failures. *Long-Range Planning*, **41**(2), pp. 177–196, 2008.

[7] Combe, I.A. & Carrington, D.J., Leaders' shifts in attention during an organizational crisis. *Leader Thinking Skills: Capacities for Contemporary Leadership*, eds. M.D. Mumford & C.A. Higgs, Routledge, 2008.

[8] Kelloway, E.K., Mullen, J. & Francis, L., Divergent effects of transformational and passive leadership on employee safety. *Journal of Occupational Health Psychology*, **11**(1), p. 76, 2006.

[9] Smith, T.D., Eldridge, F. & DeJoy, D.M., Safety-specific transformational and passive leadership influences on firefighter safety climate perceptions and safety behavior outcomes. *Safety Science*, **86**, pp. 92–97, 2016.

[10] Tafvelin, S., Hasson, H., Holmström, S. & von Thiele Schwarz, U., Are formal leaders the only ones benefitting from leadership training? A shared leadership perspective. *Journal of Leadership and Organizational Studies*, **26**(1), p. 32, 2019.

[11] Zohar, D. & Polachek, T., Discourse-based intervention for modifying supervisory communication as leverage for safety climate and performance improvement: A randomized field study. *Journal of Applied Psychology*, **99**(1), p. 113, 2014.

[12] Tafvelin, S. & Hasson, H., Agreement of safety climate: Does it affect employees' and managers' health and work performance? *Journal of Occupational and Environmental Medicine*, **61**(4), pp. e125–e131, 2019.

[13] Pahkin, K., Staying well in an unstable world of work: Prospective cohort study of the determinants of employee well-being. Finnish Institute of Occupational Health, People and Work Research Reports 107, 2015.

[14] Pahkin, K., Väänänen, A., Koskinen, A., Bergbom, B. & Kouvonen, A., Organizational change and employees' mental health: The protective role of sense of coherence. *Journal of Occupational and Environmental Medicine,* **53**(2), pp. 118–123, 2011.

[15] Ulvenblad, P. & Barth, H., Liability of smallness in SMEs: Using co-creation as a method for the "fuzzy front end" of HRM practices in the forest industry. *Scand. J. of Management*, **37**(3), 2021.

[16] Haapakoski, M., Kankainen, A. & Sjögren, T., Sickness absence at work and supporting being present at work, among employees working different shifts in the forest industry. *Journal of Ergonomics*, **5**, 2015.

[17] Richter, K., Acker, J., Adam, S. & Niklewski, G., Prevention of fatigue and insomnia in shift workers: A review of non-pharmacological measures. *EPMA Journal*, **7**(1), p. 16, 2016.

[18] Lilley, R., Feyer, A.M., Kirk, P. & Gander, P., A survey of forest workers in New Zealand: Do hours of work, rest, and recovery play a role in accidents and injury? *Journal of Safety Research*, **33**(1), pp. 53–71, 2002.

[19] Salminen, S., Long working hours and shift work as risk factors for occupational injury. *The Ergonomics Open Journal*, **9**(1), 2016.

[20] Gamboa Madeira, S., Fernandes, C., Paiva, T., Santos Moreira, C. & Caldeira, D., The impact of different types of shift work on blood pressure and hypertension: A systematic review and meta-analysis. *International Journal of Environmental Research and Public Health*, **18**(13), p. 6738, 2021.

[21] Eriksson, H.P., Söderberg, M., Neitzel, R.L., Torén, K. & Andersson, E., Cardiovascular mortality in a Swedish cohort of female industrial workers exposed to noise and shift work. *International Archives of Occupational and Environmental Health*, **94**(2), pp. 285–293, 2021.

[22] Karlsson, B., Alfredsson, L., Knutsson, A., Andersson, E. & Torén, K., Total mortality and cause-specific mortality of Swedish shift-and dayworkers in the pulp and paper industry in 1952–2001. *Scandinavian Journal of Work, Environment and Health*, pp. 30–35, 2005.

[23] Mylek, M.R. & Schirmer, J., Beyond physical health and safety: Supporting the wellbeing of workers employed in the forest industry. *Forestry: An International Journal of Forest Research*, **88**(4), pp. 391–406, 2015.

[24] Skogsindustrierna, https://www.skogsindustrierna.se/om-skogsindustrin/bransch-statistik/sveriges-och-varldens-skogar/. Accessed on: 22 Jun. 2020.

[25] Arbetsmiljölagen, Arbetsmiljölagen och dess förordning med kommentarer 21 April 2018. Arbetsmiljöverket, Sverige, 2018.

[26] Arbetstidslagen, https://www.av.se/arbetsmiljoarbete-och-inspektioner/lagar-och-regler-om-arbetsmiljo/arbetstidslagen/. Accessed on 22 Jun. 2020.

[27] Organisatorisk och social arbetsmiljö, AFS 2015:4. https://www.av.se/filer/publikationer/foreskrifter/organisatorisk-och-social-arbetsmiljo-foreskrifter-afs2015_4.pdf. Accessed on: 22 Jun. 2020.

[28] Prahalad, C. & Ramaswamy, V., Co-creation experiences: The next practice in value creation. *Journal of Interactive Marketing,* **18**(3), pp. 5–14, 2004. DOI: 10.1002/dir.20015.

[29] Grönroos, C. & Voima, P., Critical service logic: Making sense of value creation and co-creation. *Journal of the Academy of Marketing Science*, **41**(2), pp. 133–150, 2013. DOI: 10.1007/s11747-012-0308-3.
[30] Nieveen, N. & Folmer, E., Formative evaluation in educational design research. *Design Research,* **153**, pp. 152–169, 2013.
[31] Chevalier, J.M. & Buckles, D.J., *Participatory Action Research: Theory and Methods for Engaged Inquiry*, Routledge, 2013.

SAFETY ASPECTS RELATED TO IMPLANTED BREAST PROSTHESES

MICHELA ARNOLDI[1], LUCA DI LANDRO[1], GERARDUS JANSZEN[1],
MARCO KLINGER[2] & VALERIANO VINCI[3,4]
[1]Politecnico di Milano, Italy
[2]Department of Medical Biotechnology and Translational Medicine, Università degli Studi di Milano, Italy
[3]Department of Biomedical Sciences, Humanitas University, Italy
[4]IRCCS Humanitas Research Hospital, Italy

ABSTRACT

Silicone implants are largely used both in aesthetic and reconstructive medicine, thanks to their recognized biocompatibility. The damage of breast prostheses as a consequence of accidental impacts, is however one of the possible critical situations in their use. The behaviour of silicone breast prostheses in case of accidental impacts, for example in a car accident, is studied by experimental crash tests, accounting for the action exerted by seat belts during impact. Possible effects of aging over the mechanical response of the implants are also considered and approached. Static compression tests performed after long term aging indicate a possible degradation of prosthesis mechanical performance, which should be considered when analysing the possible consequences of an accidental damage.
Keywords: breast prostheses, mechanical impact, car accident, crash test, aging, damage.

1 INTRODUCTION

Silicone gel breast implants have been used worldwide for over 50 years, since their invention by Cronin in the early 1960s [1]. The number of women undergoing breast insertion, either for aesthetical reasons and for post mastectomy surgery, has steadily grown throughout the years; in 2018 more than 1.8 million implants were inserted for augmentation only and it has been estimated that 5 to 10 million women have received breast implants worldwide. The market of breast prostheses has undergone in the years a somewhat oscillatory trend, often as a result of discussions over the safety of silicone implants after long term use (e.g., activation of breast implant associated – anaplastic large cell lymphoma, BIA-ALCL) and/or in allegedly critical conditions (e.g., use of airplane or scuba diving activity) [2], [3]. Among the multiple potential complications associated with the use of breast implants, this research focuses on the implant rupture as consequence of mechanical impacts as, for instance, car accidents. In particular, the possibly incorrect use of safety belts may enhance the probability of implant rupture due to mechanical overstress in case of accident.

Data regarding the incidence of rupture rates are published by the three-major silicone gel breast implant producers (Allergan, Mentor and Sientra) [4], [5], but today there are still only limited studies that can give us an idea about the overall, actual incidence of breast implant rupture causes [6], [7]. It is especially difficult to render comparable rupture rates among different types of breast implants, manufacturers, and rupture detection methods. Among the different mechanisms proposed for implant rupture, the most common is the damage from surgical instruments (50%–64% of all causes), followed by unidentified opening/rent without indication of cause (no evidence of sharp instrument damage, 35%–37%), fold flaw (8%), shell swelling (reduction of the shell strength due to migration of the silicone fluid from the gel to the shell), delamination, manufacturing defect, surgical impact and trauma to the implant, like an external pressure to the chest or closed capsulotomy. Focusing on the traumatic rupture, it is reported that most trauma do not cause a prosthetic rupture, except in case of impact with sharp objects. On the other hand, it recognized that many rupture can

WIT Transactions on The Built Environment, Vol 214, © 2022 WIT Press
www.witpress.com, ISSN 1743-3509 (on-line)
doi:10.2495/SSR220151

remain silent for long time with, without evident sign for the patient [6]. An accurate review of the scientific literature shows that there are no studies regarding the relationship between the impact loads, which typically can occur in a car accident, and the risk of breast implants rupture. Although there is a wealth of scientific papers related to the medical consequences of implant rupture and to the clinical methods of detecting damages [7]–[9], engineering-based investigations about the possible conditions leading to rupture are very limited [10].

As a matter of fact, in the event of a car accident, loads from the compression exerted by the seat belt may be transferred to the breast. Of course, even worse is the case of accident with no seat belt used, leading to direct impact with steering wheel or other parts of the vehicle. For this reason, our group approached a preliminary study at the mechanisms and the physical aspects of the impact and how this is related to the possible damages of the breast implants. It is important to stress that, as evidenced by the cited papers, a damage of the implant following a car accident does not necessarily require surgical intervention, but in many cases a visit by a physician and/or an instrumental check is highly recommended.

A discussion about the danger of rupture should consider also the possible aging state of the prosthesis at the time of the accident. Silicon rubber employed in breast prostheses is selected also on the basis of its long-term properties, which guarantee a long endurance to the implants. The duration of an implanted prosthesis is estimated in the range of 12 to more than 35 years [11]; moreover, it should be considered that in most cases, removal or replacement is the result of causes not related to the implant aging or defects. However, from the clinical point of view, the aging is usually intended in terms of swelling or diffusion of internal gel, or in terms of biomedical effects over the surrounding tissue. Such effects are object of common testing procedures by producers, as required by national and international standards and regulations, e.g., EN ISO 14607:2018. On the other hand, from a mechanical point of view, the aging of cross-linked shell rubber may lead to mechanical stiffening and reduced deformability, which can result in degradation of mechanical performances in case of accidental impacts. In order to estimate the possible relevance of such phenomena, mechanical tests after periods of permanence in biological saline solution at the body temperature were performed.

For the explanted implants, a sensitivity on aging was presented, although no correlation between aging, dimensions and shape of the implant could be defined at present, since previous history of the patients and implants could not be retrieved in detail. Further investigations with both new and explanted implants are being performed to better evidence and estimate possible aging effects.

2 EXPERIMENTAL

Prostheses explanted from living patients for medical reasons, not related to previous mechanical impacts or damages, were selected for the tests. Two different shapes (round and raindrop implants) were selected, which were produced by a primary breast prostheses manufacturer. The round and raindrop implants had a volume of 150 cm^3 and 180 cm^3 respectively.

Compression tests were carried out to determine rupture loads and deformations. A maximum pressure limit was identified after which the implant developed a permanent deformation and was more prone to rupture. Tests were carried out with a MTS858 Mini Bionix II testing machine, by pressing the prostheses between two flat metal plates at constant, slow crosshead rate (1 mm/min). Squeezing deformation at failure and internal pressure were estimated during the tests. The internal pressure was estimated as the ratio between applied force and measured footprint of the deformed implants just before the rupture. Data from these tests were used for a preliminary identification of possible critical conditions for subsequent static test after aging and for dynamic impact tests.

To estimate possible effects of aging of shell rubber, prostheses were maintained in 0.9% NaCl saline solution at 40°C for about one year and three months, when rupture occurred during tests. DSC and dynamic-mechanical analysis of aged shell material were performed during this time span which, however, could not evidence detectable changes in the glass transition or other thermal properties. Static compression tests were carried out on a prosthesis (raindrop shape) aged for different times, until rupture occurred. The specimen was periodically compression tested at a force, which was 70% of the rupture compression load initially measured on an equal prosthesis.

Dynamic tests simulating a car accident were carried out with a specific instrumentation commonly employed to test the effects of structural crash impacts and, in particular, to assess the safety of seat belts. The test facility consists of a sled that can be moved along a 80 m long rail, to a desired speed and stopped with a selected deceleration program. Deceleration is controlled by a set of plastic deformable aluminium beams connected to the sled, which undergo extensive buckling during impact.

An anthropomorphic dummy (Hybrid II; H2-50) with a round breast prosthesis applied, was placed on a seat with belts, which was mounted on the sled for the tests. Accelerometers placed on the sled and directly connected to the dummy allowed to record the actual deceleration history. Visual inspection of breast prostheses was done to evidence failure signs after deceleration tests, in particular to detect possible leakage of internal gel.

3 RESULTS AND DISCUSSION

Static tests were carried out on two different kinds of prosthesis (round and raindrop shape) and evidenced a limit of the static load and pressure that these can sustain. Fig. 1 shows the round prosthesis before and after complete rupture by compression.

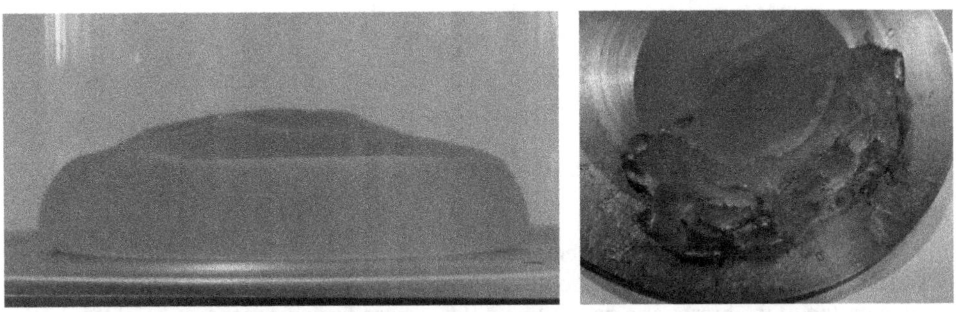

Figure 1: Round prosthesis before and after rupture in compression test.

Fig. 2 shows the compression set-up before and during tests. The load, plates positions and contact area between deformed prosthesis and plates was recorded throughout the test. The results of compression tests are compared and reported in Fig. 3.

Rupture occurred at about 3 and 4 kN of pressing force for round and raindrop implants. Considering that the shell material is the same for the two implant shapes, it is not surprising that in both cases the rupture starts at a very similar internal pressure, of about 2 bar. It is to be noted that the estimated internal pressure at rupture is not at all related to the limit hydrostatic pressure that the implant can sustain, for example during a scuba diving

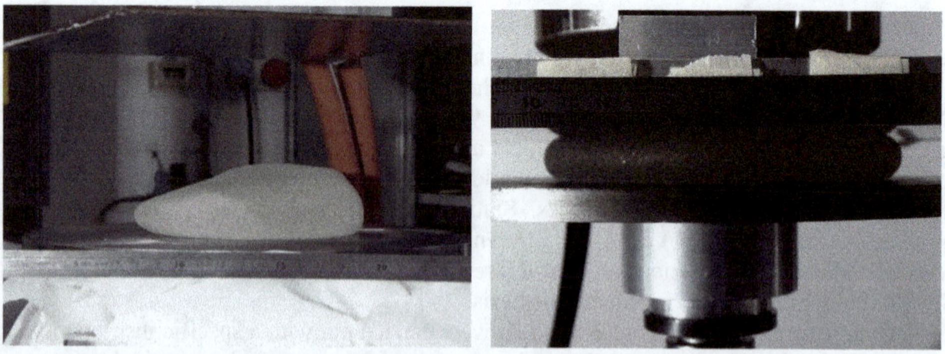

Figure 2: Compression test set-up of rain drop shape prosthesis.

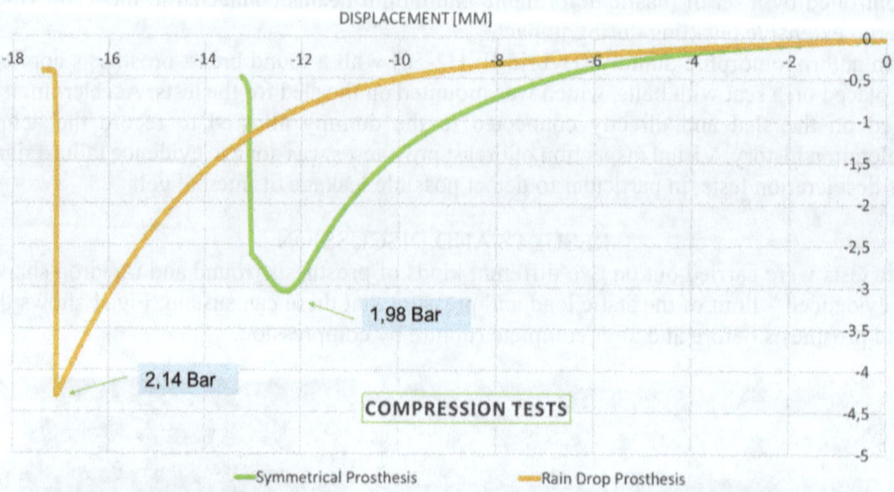

Figure 3: Compression test results of two different shaped prostheses.

immersion or in a hyperbaric chamber. However, it can be considered relevant in a number of situations of impacts against blunt objects, as for example a car accident with a safety belt passing on top of the breast or a bad fall on a flat surface from a consistent height. It is also to be noted that such impacts can result substantially more critical than those indicated for the testing of new prostheses as required by EN ISO 14607:2018. The results were also relevant in the selection of impact velocity and maximum deceleration in case of vehicle impact tests later preformed.

To assess possible influence of ageing over shell rupture elongation and, in general, on the overall behaviour of the prostheses, static tests were repeated with the same loading program, after selected aging periods (few months) up to more than one year. In these tests the specimen pressing was stopped at 70% of the rupture load measured before aging; this corresponds to an inner pressure slightly lower than 70% of rupture pressure. No visual signs of material changes were detected after tests with implant aged up to about one year. However, after 15 months aging, rupture with internal gel leaking was evidenced. This result suggests that a significant physical modification of shell properties may occur with aging time. Considering the very limited number of loading cycles before rupture, possible effects of repeated tests are not deemed relevant.

Dynamic impacts were tested to simulate car accidents at different vehicle speed. Breast implants were mounted on an anthropomorphic dummy by adhesive tape, positioning a safety belt passing over the prosthesis. This may appear an incorrect position, but breast damages due to the pressure of seat belts are certainly not unusual [12]. Fig. 4 shows the dummy assembly for the tests.

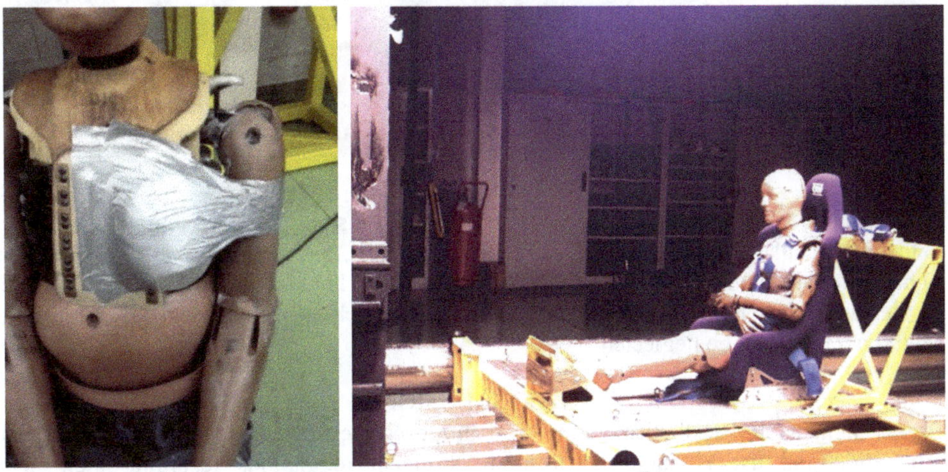

(a) (b)

Figure 4: (a) Prosthesis implant fixed on Hybrid II dummy; and (b) Crash test facility.

Two crash conditions, simulating impacts at 50 km/h and 90 km/h were selected. In the first situation, no evident rupture was visually observed. In the second crash conditions a clear rupture, with loss of gel, was observed (Fig. 5). The measured maximum deceleration was of about 25 g for both impacts, but in the second crash the time span was of course longer. Fig. 6 shows two instant frames of the impact events recording.

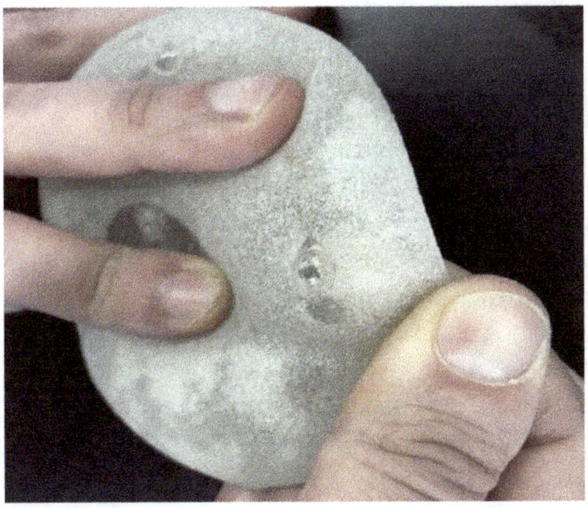

Figure 5: Damage of the prosthesis and gel leakage after impact.

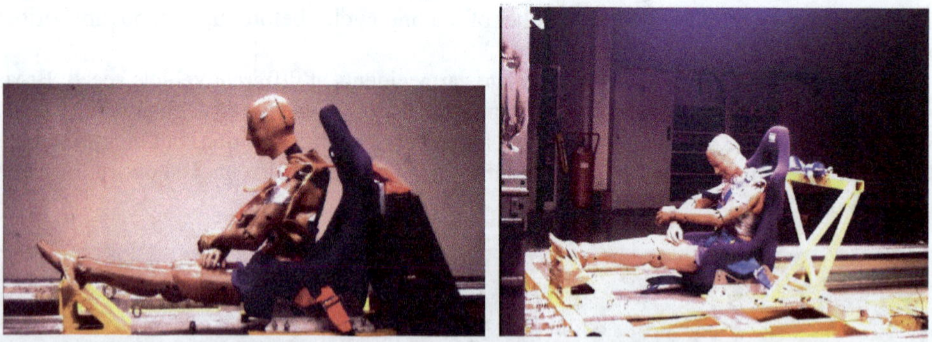

Figure 6: Instant pictures of the dummy during impact.

It should be noted that car accidents in such conditions may produce important damages to the car occupants; safety features, i.e. belts and airbags, are intended to be highly efficient right in these situations, to avoid primary consequences [13]. On the other hand, the chance of "secondary" or minor consequences of impacts, which are not immediately detectable, should not therefore be neglected.

4 CONCLUSIONS

Breast implants for aesthetical or medical purposes are nowadays very common. Their production and use is subjected to quite stringent regulations regarding their body compatibility and durability. On the other hand, there are situations which may require additional attention, related to a safer use of such prostheses. In particular, in case of relatively high energy impacts as those involved in car accidents or similar impacts, prostheses may undergo to damages not immediately detectable, which, on the other hand, may produce delayed effects. In addition, the long-term use of such implants may lead to ageing of the physical and mechanical properties that, although little relevant in terms of body compatibility or reaction, can result in a variation of mechanical performances and response during potentially critical impacts. In this paper, some evidence of such behaviour is reported, which strongly suggests a better attention to such aspects and more thorough investigation about possible long term aging effects of implanted prostheses.

ACKNOWLEDGEMENT

The financial support of Project MUR-PRIN 2020, Prot. 2020LYY27Z, "Breast implants safety: is our knowledge enough?" is gratefully acknowledged.

REFERENCES

[1] Cronin, T.D. & Gerow F.J., Augmentation mammoplasty: A new "natural feel" prosthesis. *Transactions of the Third International Congress of Plastic Surgery*, Excerpta Medical: Amsterdam, pp. 41–49, 1964.

[2] Jalalabadi, F., Doval, A.F., Neese, V., Andrews, E. & Spiegel, A.J., Breast implant utilization trends in USA versus Europe and the impact of BIA-ALCL publications. *Plastic and Reconstructive Surgery – Global Open*, **9**(3), e3449, 2021. DOI: 10.1097/GOX.0000000000003449.

[3] EU, Scientific Committee on Health, Environmental and Emerging Risks (SCHEER), Final opinion on the safety of breast implants in relation to anaplastic large cell lymphoma, 26 March 2021. https://health.ec.europa.eu/system/files/2021-04/scheer_o_018_0.pdf.

[4] Hammond, D.C., Migliori, M.M., Caplin, D.A., Garcia, M.E. & Phillips, C.A., Mentor contour profile gel implants: Clinical outcomes at 6 years. *Plastic and Reconstructive Surgery,* **129**, pp. 1381–1391, 2012. DOI: 10.1097/PRS.0b013e31824ecbf0.

[5] Spear, S.L. & Murphy, D.K., Allergan silicone breast implant U.S. core clinical study group. Natrelle round silicone breast implants: Core Study results at 10 years. *Plastic and Reconstructive Surgery,* **133**, pp. 1354–1361, 2014. DOI: 10.1097/PRS.0000000000000021.

[6] Hillard, C., Fowler, J.D., Barta, R. & Cunningham, B., Silicone breast implant rupture: A review. *Gland Surgery*, **6**(2), pp. 163–168, 2017. DOI: 10.21037/gs.2016.09.12.

[7] Handel, N., Garcia, M.E. & Wixtrom, R., Breast implant rupture: Causes, incidence, clinical impact, and management. *Plastic and Reconstructive Surgery,* **132**, pp. 1128–1137, 2013. DOI: 10.1097/PRS.0b013e3182a4c243.

[8] Glazebrook, K.N., Doerge, S., Leng, S., Drees, T.A., Hunt, K.N., Zingula, S.N., Pruthi, S., Geske, J.R., Carter, R.E., McCollough, C.H. & Fletcher, J.G., Ability of dual-energy CT to detect silicone gel breast implant rupture and nodal silicone spread. *American Journal of Roentgenology*, **212**(4), pp. 933–942, 2019. DOI: 10.2214/AJR.18.20138.

[9] Bengtson, B.P. & Eaves, F.F. 3rd., High-resolution ultrasound in the detection of silicone gel breast implant shell failure: Background, in vitro studies, and early clinical results. *Aesthetic Surgery Journal*, **32**(2), pp. 157–174, 2012. DOI: 10.1177/1090820X11434507.

[10] Janszen, G., Vinci, V. & Klinger, M., Mechanical characterization of breast prosthesis: Static analysis. Presented at *3rd Global Conference and Expo on Applied Science, Engineering and Technology and 6th Global Congress and Expo on Materials Science and Nanoscience*, Dubai, UAE, 2019.

[11] Prasad, K., Zhou, R., Zhou, R., Schuessler, D., Ostrikov, K.K. & Bazaka, K., Cosmetic reconstruction in breast cancer patients: Opportunities for nanocomposite materials. *Acta Biomaterialia*. **86**, pp. 41–65, 2019. DOI: 10.1016/j.actbio.2018.12.024.

[12] Song, C.T., Teo, I. & Song, C., Systematic review of seat-belt trauma to the female breast: A new diagnosis and management classification. *Journal of Plastic, Reconstructive and Aesthetic Surgery,* **68**(3), pp. 382–389, 2015. DOI: 10.1016/j.bjps.2014.12.005.

[13] Pipkorn, B. & Kullgren, A., Effects of active structures on injuries in medium severity frontal impacts. *Proceedings of the 21st (ESV) International Technical Conference on the Enhanced Safety of Vehicles*, Paper No. 09-O380, Stuttgart, Germany, 2009.

Author index

WIT PRESS ...for scientists by scientists

Urban Water Systems & Floods IV

[illegible]

[illegible]

[illegible]

[illegible]

[illegible]

[illegible]

[illegible]

www.ingramcontent.com/pod-product-compliance
Lightning Source LLC
LaVergne TN
LVHW062238200926
848484LV00006B/1570

9781784664831